# Fire, Ice
## and Paradise

by

### H. Leighton Steward

*AuthorHouse™*
*1663 Liberty Drive, Suite 200*
*Bloomington, IN 47403*
*www.authorhouse.com*
*Phone: 1-800-839-8640*

*First published by AuthorHouse   6/13/2008*


*ISBN: 978-1-4343-8999-2 (sc)*


*For additional copies contact AuthorHouse Book Orders at : 1-888-280-7715*

*Printed in the United States of America*
*Bloomington, Indiana*

*This book is printed on acid-free paper.*

# TABLE OF CONTENTS

# FOREWORD

Leighton Steward has a history of taking issues of seemingly considerable complexity and boiling them down to a few salient points that leave his readers more knowledgeable and enlightened. The task of reducing global climate change, which includes global warming, to a few key points for his intended audience, the layman, is particularly challenging, but I believe Leighton has accomplished it. In addition, he has included several summary tables and charts that will allow reader to follow the ever-evolving research results on past and present climate and will stimulate their curiosity and expand their knowledge of Earth's dynamic history. He chronicles that Earth was birthed in fire, almost frozen in ice and is now in an under-appreciated paradise of a climate.

Growing up outside a small town in a rural setting, Leighton had daily exposure to the fields and the woods and the animals that lived therein. His affinity for this setting was probably what gave him an early awareness of Earth's more pristine environments. As the years sped by, Leighton left his home town and concrete replaced the grass, and skyscrapers supplanted the natural shade from the trees, but he obviously never lost his appreciation for his earlier environment.

He pursued a geology degree at Southern Methodist University and minored in geography, which introduced him to meteorology. Meteorology is a study of the weather and the factors that not only cause it but also cause it to change. His early interest in the climate has developed into a serious fascination with and study of Earth's earlier climates, its paleoclimates.

The paleoclimate interest came naturally in other ways. A master of science degree in geology provided him with the knowledge of other Earth processes, both present and past. As you will see, many of these natural processes affect the climate. The climate had a profound effect on another of his early interests, the environment.

Following college, as reconnaissance leader of a photo-mapping squadron when he was an officer in the U.S. Air Force, Leighton spent a significant part of three years in the Amazon and Orinoco basins of South America, traveling by foot, dugout canoe, burro, Jeep, or the occasional helicopter. What he witnessed was pristine; no clear-cutting of the forests - no cutting at all. Like most of us, he doesn't like what he sees happening there today.

As good fortune would have it, he, later became head of the Louisiana Land and Exploration Company, an energy company that was also the largest private owner of coastal wetlands in the lower 48 states – 600,000 acres in all. Located in the Mississippi River delta this fragile land is still under siege, fraught with the problems of land loss caused primarily by man's leveeing the huge river and starving the delta of new sediment. Leighton's first crack at authorship, with yeoman help from his wetlands manager, Bill Berry, resulted in a publication that explained exactly why the land loss was occurring; a true accounting that had not been picked up by any of the media. First published in 1988, *Louisiana's National Treasure* was a success and changed both the media's characterization and the public's opinion of why the wetlands were disappearing. If a lawmaker doesn't know what is causing a problem, how can he or she enact legislation to properly fix that problem? More than 110,000 copies of this publication now reside in high schools, universities, and governmental agencies, and in the hands of other interested parties. I frequently hear that the secret of *Louisiana's National Treasures'* success is that he simplified it, made it easy to understand, and did not propagandize for any particular interest.

Because of his company's voluntary efforts in trying to save this huge wetlands system, as well as his recommending measures that would benefit all of the nation's wetlands, the Environmental Protection Agency presented his company and him with it's highest honor for environmental excellence.

Next came Leighton's work as the lead author of a book on how people should live and eat to lose weight and avoid debilitating illnesses, particularly diabetes. After sixty-two drafts and sales of 210,000 copies of a self-published edition, he and his coauthors turned it over to a national publisher, and the book *Sugar Busters: Cut Sugar To Trim Fat*, became an instant number one *New York Times* bestseller. More than four million copies of *Sugar Busters* and its related offspring have made it to the printing press. Interestingly, Leighton received the same comments as he did with *Louisiana's National Treasure*. The subject matter was presented in simple, easily understandable language. It was effective, and has been imitated by other authors and diet companies as well as the food manufacturing industry that now offers consumers the choice of whole grain, high fiber and no sugar added products.

Leighton's early career included a stint as a field leader for a project in the Gulf of Mexico, drilling and coring sediments below the sea floor, taking water samples, and studying surface to sea floor current conditions on a truly ground breaking research project on the upper continental slope in 600 to 4,000 feet of water. This was from 1964 to 1966, when such a coring project had never been done in the Gulf. It taught Shell, his employer, about sediment and fossil types deposited far beyond the current coastline, those deposited nearby when sea level was low during the coldest parts of the glacial cycles and then the return to the present high sea level stand of this warm, interglacial period.

Years later, in studying the reasons for the loss of the lower Mississippi River wetlands system, the specter of climate change and rising sea level again became a key area of concern. For ten years, Leighton was chairman of the *National Wetlands Coalition*, and for five years chairman of the *Audubon Nature Institute*, where nearly every recommendation or decision was weighed against a background of its effect on the pertinent ecosystem. He was often invited to give presentations on how to have a net gain in our country's wetlands, as well as other topics of national interest, such as finding a better way to obtain oil to fill the *Strategic Petroleum Reserve* and changing our government's nutritional guidelines. One-on-one presentations and recommendations were given to two EPA administrators, two secretaries of the Department of the Interior, two secretaries of the Department of Energy, and several senatorial and congressional committees. In addition, he was asked to be the spokesman for the independent sector of the energy industry on **_presidential missions_** to the then USSR and Turkey (for G. H. Bush) and Pakistan (for Clinton). Obviously, someone trusted his advice!

Leighton received several awards from the energy industry, for his leadership in proactively and voluntarily spending (non-requisite) dollars and efforts in environmental areas. The most prestigious energy industry trade association, The American Petroleum Institute, (API), presented him with their highest award, the Gold Medal for Distinguished Service, for being an exemplary leader in environmental practices.

With his educational background, lifelong interest in the environment, and an entire world wallowing in confusion about what is causing our Earth's climate to change, Leighton thought, why not "go for three"? Why can't global climate change be explained as clearly and simply as wetlands loss and the best diet for mankind? This is an extremely complex subject and not easily reducible to two or three one-liners as to what makes Earth's climate change. But who better than Leighton to do it, otherwise we may have to settle for an overly scientific version written by a well-meaning PhD who may be an expert on the subject, but is so brilliant that most of the writing or presentation goes right over the nonscientists' heads.

I think Leighton is eminently qualified to write an objective, layman-level review of the complex interaction of factors that come together to cause global climate change.

Gene Shinn, PhD
Courtesy Professor
Marine Science Center
University of South Florida

# PREFACE

*Fire, Ice and Paradise* is meant to be an educational and interesting guide to global climate change. The intended audience includes the laymen and policymakers who need fact-based input to help them better understand the climate-change issues. Most media information is only someone's opinion or a sensationalized soundbite and not based on scientific fact or even best current science. The many summary tables and charts that are presented here are intended to provide everyone with a handy reference to look to as the climate-change issue continues to evolve.

The initial inspiration for writing this book came from a discussion with Dr. Crayton Yapp, professor of geological sciences, at my alma mater, Southern Methodist University. Dr. Yapp's research was indicating that the Earth's carbon dioxide ($CO_2$) levels were probably as much as 18 times higher 500 million years ago, during the Cambro-Ordovician Period, than the $CO_2$ levels of today. At the time, I was unaware that Earth's paleo (old) atmosphere had contained such high levels of this greenhouse gas.

I then began pouring through books, textbooks, and articles written by other experts in climate study and was further inspired by discussions with other professors at SMU and by remembering conversations or presentations on factors that cause climate change while I was a member of the Advisory Board at *Lamont Doherty Earth Observatory* at Columbia University. Lamont had been the first institution to have great predictive success relative to future El Niño - La Niña Pacific Ocean circulation events, as well as acquiring the magnetic data and mapping the ocean floor to confirm the continental drift or sea-floor spreading hypothesis.

Summarized here for the general public are the factors that cause or "drive" the climate as well as discussions of the frequent climate changes that have occurred and will continue to occur on Earth. These factors are numerous and their interactions can be very complex or, in some cases, fairly obvious and straightforward. I will attempt to limit most of my observations to the major generalizations that can help the reader get a better feel for what is going on and why. From time to time, examples will be cited of the complexities of climate studies that can drive even the best researchers and experts crazy as they attempt to understand the past, much less make reliable predictions about the future.

The information presented here should help you better understand and interpret what you hear every day and also enjoy or connect with it more. More knowledge will help screen the often sensationalized media spins injected into otherwise straightforward stories or facts regarding the climate.

In the figures and charts, I will often use the abbreviations *bya* for billion of years ago, *mya* for millions of years ago, and *kya* for thousands of years ago, and *ppm* for parts per million. *GHG* will be used for greenhouse gases and $CO_2$ for carbon dioxide, $O_2$ for oxygen, $CH_4$ for methane, and *E.T.* for extraterrestrial. Temperatures will be identified in either degrees Fahrenheit (*F*) or Celsius (*C*) and I have provided a comparative scale so you can translate the temperatures mentioned here or those found in other publications into the scale in which you are most comfortable (Figure 1). Distances are often given in miles or kilometers. One kilometer is approximately 0.6 mile or a mile is approximately 1.6 kilometers.

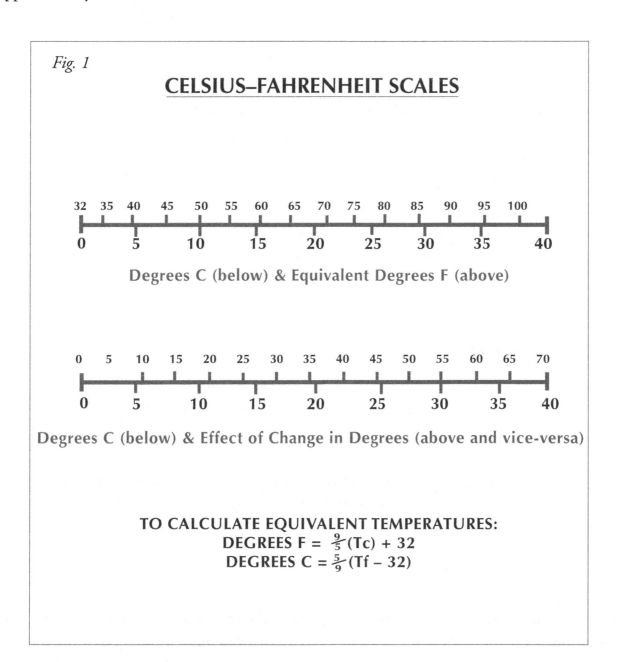

*Fig. 1*

## CELSIUS–FAHRENHEIT SCALES

Degrees C (below) & Equivalent Degrees F (above)

Degrees C (below) & Effect of Change in Degrees (above and vice-versa)

**TO CALCULATE EQUIVALENT TEMPERATURES:**
**DEGREES F = $\frac{9}{5}$(Tc) + 32**
**DEGREES C = $\frac{5}{9}$(Tf − 32)**

To fully appreciate the pace of Earth's historical evolution, I ask you to try and comprehend very, very long periods of time and very, very slow Earth dynamics, such as continental movements that are measured in one to three inches a year (Figure 2). For instance, one inch a year is a foot in 12 years, a mile in 63,360 years, and 100 miles in 6.3 million years. The current width of the Atlantic Ocean has been created in about 175 millions years, as the Americas split and then drifted away from Europe and Africa.

*Fig. 2*

# TO UNDERSTAND THIS SUBJECT BEST

## THINK IN VERY LONG PERIODS OF TIME

100 years ago — First airplanes and cars; pneumonia meant death

1,000 years ago — Medieval times, warmer than today

10,000 years ago — Earth warmed to current level from ice age

100,000 years ago — The last time Earth had warmed

1,000,000 years ago — Earth has had 10 ice ages since then

10,000,000 years ago — No permanant ice in Northern Hemisphere

100,000,000 years ago — Earth was 20° F warmer, $CO_2$ eight times higher

1,000,000,000 years ago — Was 1,000 million years ago!

4,600,000,000 years ago — Earth was formed

## THINK SLOW

Earth's continents are drifting at one to three inches per year. That is one foot in 12 years, one mile in 63,360 years, and 100 miles in 6.3 million years.

The shape of Earth's orbit around the sun changes from nearly circular to most elliptical and back every 100,000 years.

*H. Leighton Steward, 2007*

Relative to the amounts of the various gases that make up our atmosphere, I was advised by a former member of the United States senate, when acknowledging that very few of our lawmakers have any significant background in science, that they are going to vote on technical issues in the way they or their constituents understand the issues. He continued that they certainly didn't know of what significance was a part per million or part per billion. Let's start with one part per hundred which is the same ratio of one penny to one dollar (100 pennies). Following that start, one penny per 1,000 pennies (10 dollars) is still comprehensible. One part per million is one thousand thousand and one part per billion is one thousand million or a very small speck in the atmosphere. Whether that tiny speck is of significance depends on the impact it can impart. Many substances that are deadly at parts per hundred are totally benign at parts per million or billion. Arsenic, a naturally occurring "poison" in Earth's environment, is a good example. At a few parts per million or billion of arsenic it has no poisonous effect on human life.

Without the training to know how many parts per whatever are significant, the lawmaker is still at the mercy of someone who can tell him or her what level is significant. If I wanted to stop the development of a facility that might release one part per billion of arsenic into a stream, I would just feed the media the arsenic word and scare the hell out of the voters and lawmakers and then sit back and watch the experts try and dig themselves out of the hole that had been created by my simply floating the arsenic word.

Where does that leave you, the non-scientist reader? Unfortunately it leaves you at the mercy of the party feeding you the information. My best advice, when you hear of a coming catastrophe, is to get a second, non-biased opinion and ask them for some peer reviewed evidence to back up their position. This is particularly important when you hear only a one-liner summary of an impending catastrophe.

I sincerely hope you receive lasting value from reading this book, whether or not you agree with the conclusions I put forth. The opinions contained in this book are my own and should not be misconstrued as those of anyone I have identified in various ways in my text and acknowledgements.

So sit back, adjust the light just right, and allow me to share some brief remarks and a liberal use of summary charts and illustrations to see if we can reduce the key points about Earth's past and present climate to as few as ten. Review them while considering the scientific evidence or expert opinions contained in the text or illustrations.

# INTRODUCTION

This book is a biography, primarily of the Earth's climate history, summarized from research papers and works of many scientists and knowledgeable authors, plus information obtained from personal conversations with several of them. It also includes my own observations and thoughts. The audience I am targeting is _you_, whether you are a layperson or a lay scientist. If I cannot get you interested in global climate research, I will have to settle for making you the best informed person on the block regarding Earth's past and present climate. In your lifetime, this topic is not going to go away. If anything, it will become more and more a part of everyday conversation.

How the world addresses global climate change will be one of the most important topics of our lives. The many factors that determine climate change are driven by the various laws of science. The policies that are put into man's laws are determined by legions of nonscientists who are naive regarding what is or is not causing climate change. Included in this group are presidential candidates, one of whom will become extremely important in establishing the future policies and laws. Thus the need to get some "soft" science circulated to these key people as well as the general public that elects them. Otherwise, we may see a misdirection of our resources that will be spent dealing with only one narrow issue concerning global climate change.

I will discuss the major changes in Earth's climate, atmosphere, plant and animal life, and overall environment. The review begins a few billion years ago and continues right up to today. Relying on what is being reported in reputable scientific journals, university textbooks, and personal communications with well-known climate researchers, I also will tell you about the continuous small changes of more recent times to alert you to the fact that global climate change (GCC) is constant and that global climate stability is really a rare condition. Whether during the relatively warm intervals such as what we are experiencing today, or during the extended glacial cycles, climate has jumped up and down frequently, with many of the large changes happening over the course of only a few years or decades.

These constant climate changes, what we scientists call "chatter" or frequent changes which in many cases altered Earth's average temperature only a few degrees Fahrenheit, were sufficient to alter the course of many past civilizations. Examples will be given to substantiate that hypothesis. As we go farther back in time, the curves indicating greenhouse gas content and temperature are averaged (or "smoothed") because of the current sparsity of data and/or analyses. As additional data pours in, you'll see that more climate variation or chatter will show up and wrinkle these smoothed curves.

Near the end of this book, I will recommend other writings that will be of great interest to those who want to know more about climate history and also about anthropology (the study of the history of humans) and how climate affected some of our predecessors.

This book contains a lot of science but, is written in layman's language as much as is possible. While I have attempted to summarize the better-known facts, if I should mention some less solid facts or speculate a little bit, I will identify my comments as such and use a "probably" or "possibly" to alert you to an as-yet-unproven point. When a proven answer does not exist, you often will get my informed opinion since I think I owe you that.

Most authors who write on subjects involving technical studies, proper nutritional guidelines, or, particularly, in voicing a specific political view, claim to have the exact answer to the subject matter at hand. <u>This is not such a book!</u> While I hope this is the best current summary, if not simplification, of this extremely complex subject, all of the answers about why Earth's climate was like it was for the past 4.6 billion years do not yet exist.

Once you have digested the material presented herein, I hope you will be able to speculate on your own about how and why the climate changes or has changed and join me in looking forward to the new knowledge that is continuously becoming available in scientific literature. Trips to the post office that I used to dread for all the bills that were lying in wait now create anticipation for *Science, Nature, Geotimes*, the science sections of the newspaper; any articles that might contain a brand new discovery that can be plugged into the summary charts and tables contained in this book.

If you see a book, a film, or an article that claims to contain all of the answers to past and future climate change, be very skeptical, because no such thing is possible in our current state of knowledge. In fact, one of the goals of this book is to show that *there is no one answer*. Sometimes we are so hungry for answers that we accept certain conclusions too easily. So, again, be particularly skeptical when climate changes, such as global warming, are declared to be caused by a single factor, such as the currently popular belief that global warming is caused only by the increased $CO_2$ content in the atmosphere. $CO_2$ is only one factor in a complex system containing many active climate drivers.

I want you to get to know Earth better; its relationship to the sun and the moon, how and where its continents have moved, and particularly its climate history. Will our current paradise of a climate be preserved or become another paradise lost? I hope some of you will want to support, with your vote or influence (or even dollars), more basic research so we will be better prepared to try to extend the life of this current paradise and not see it disappear into the grips of another ice age like that which ended the earlier, comfortably warm interglacial climates within the last one million years of Earth's history.

You will learn about volcanic lava flows, volcanic explosions, extraterrestrial impacts, drought and dust storms, continental flooding, and extreme ice cover (glaciations). All of these events, which have been portrayed as examples of abnormal weather or catastrophes, in fact have always occurred quite frequently throughout time. The reason they seem more frequent and catastrophic now is because of the current 24/7 global media coverage, which was non-existent until very recently. Examination of official records from decades past shows that the frequency of such events is really no greater today. It just seems that way.

Does this imply that we have had no effect on Earth's recent climate? No. If we put anything into the atmosphere or the oceans, or alter the presence or amount of forests, grasslands, marshlands, or anything that changes the reflectivity or heat absorption of the surface of the planet, we will change the overall environment in some minor and usually insignificant way. But if we add or alter something in a larger, more far-reaching way, we may instigate changes that do have a more measurable effect. For an illustration of our growing leverage to cause an effect, see Figure 3.

Will alterations caused by us, if we should be so lucky as to develop that power, be good or bad? It depends on what it does to you or to others that inhabit the planet. This impact would include not only humans but other fauna (animal life) and flora (plant life) as well. Our ultimate goal should be not to make alterations or changes that, while temporarily advantageous, eventually could lead to very unintended consequences such as causing Earth to get colder.

What is the best thing we can do to understand changes we might make that would result in a positive change or stabilization, and, also, what might produce a negative result? We need to understand better what has occurred and where, when and why, and then make an informed judgment about what to try to do or not do. How can we accumulate this knowledge? The sediments, rocks and fossils have been waiting to tell their story of Earth's history for eons. This history includes information on Earth's climate, its inhabitants, the land and ocean's relative positions on Earth, and even evidence of "visitors" from outer space. Yes, there has been research, but scientists only recently have developed the technology and tools to decipher Earth's "code". We must go to Earth's "library" (see Chapter Two) and find out what all this new information means. With more studies of what is in Earth's "library", the data will expand and the resulting interpretations will become more precise.

If you are currently a student or teacher, I hope this will make you want to study some of the basic or peripheral sciences that apply to our climate or its history. Short of that, since the weather is the number one most-talked-about subject in conversations these days, you instantly will become one heck of a conversationalist!

If you enjoy solving puzzles, you should like this subject matter. The major causes and effects of climate change are defined and discussed in terms that will help most of you understand their meaning. However, the various causes and effects, which may be obvious and simple when analyzed separately, become much more complex when several of the pieces become active, or reactive, at the same time. Chapter Three on Climate Drivers contains a list of all the different pieces and how many of them interact. This will help you understand the reasons for the "heated" debates on what is currently causing global climate change.

Fortunately for those of us who like to solve problems, a scientific consensus on many of the answers has not yet been established. Although you may not be a true scientist yourself, by becoming knowledgeable about the individual pieces, you may just gain an interesting insight regarding potential answers. As an informed citizen, you can share your questions with the experts and legislators who may not have looked at the climate changes from your particular perspective and your input might have implications for future policy making. "More heads are often better than one."

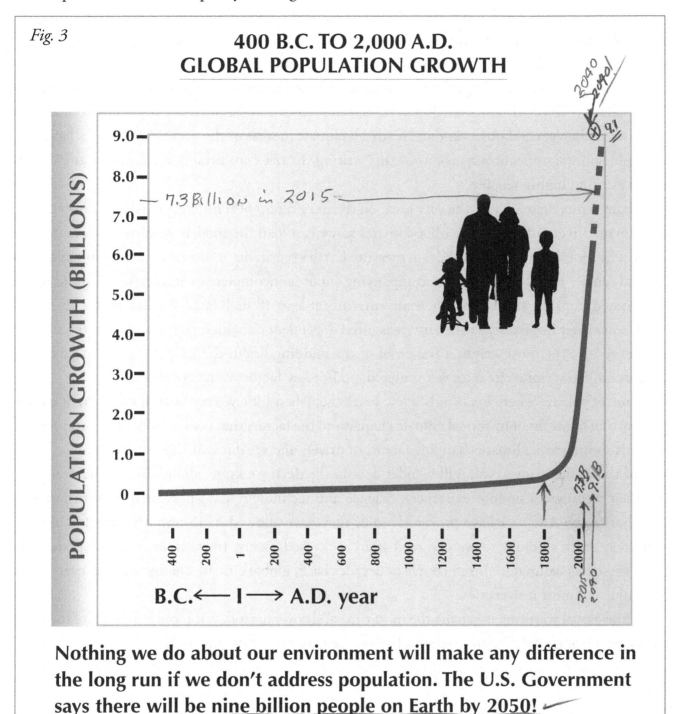

*Fig. 3*

**400 B.C. TO 2,000 A.D. GLOBAL POPULATION GROWTH**

**Nothing we do about our environment will make any difference in the long run if we don't address population. The U.S. Government says there will be <u>nine billion people on Earth by 2050</u>!**

*Gerhard, 2007*

Also, to demonstrate that there really are many sources of data that allow these conclusions to be drawn from the past, I will summarize the data sources or paleoclimate indicators for you. These indicators will range in quality from excellent to only fair or directional, and I have so labeled the limitations of each in my summary chart.

I have not included a separate write-up on the intricacies of climate models, a complex subject in itself. Models can be helpful, even important. But while the science of modeling can produce some precise, reliable results in some fields, such as predictions of stress capacities for buildings or dams, models have been less precise in the area of paleoclimatology, and for good reason. Our understanding of the historic effects of individual climate drivers is only in its early stages, not to mention our understanding of the magnitude of the effects of many of these drivers and how they interact with other climate drivers. Again, models can be helpful, but my opinion of them is summarized in a general observation. Nearly all climate models of the past three decades have been wrong in some significant way, so, as of this writing, I can't comfortably recommend any current model as being highly reliable.

We need more research (at the very least, on the most recent 500 million years) into how and why the climate drivers affected the earlier climates so we can load the models with more accurate data. When a factor is entered into a model to measure Earth's sensitivity to the factor, that can make some rational sense. But if a lot of the accompanying input is incomplete or imprecise, accurate results are doomed from the start. Also, there are currently at least 18 such factors whose effects are, at this point, somewhat imprecise, so the answer is often dependent on which factor or factors the modeler chooses to give the most weight. The proof of the pudding lies in the lack of success in predicting the past climates; imprecise data were entered, and/or key factors were left out.

Some of you are where I was only a few years ago, when I knew very little about current climate change, much less about historical climate change and the factors that controlled it. Immerse yourself into this summary of climate changing factors or drivers and see if it will also "turn you on". As you read of the ancient times, you will wonder how in the devil we know about climates before people and their writings came into existence. Science and technology have presented us with just such an opportunity. And as I have previously said, they have allowed us in only the past few decades to unlock Earth's library so we can read from rocks, sediments, trees, snow, ice, and pollen, etc. about these paleoclimates. Just a couple of decades back, global climate change was not even on the curriculum of most universities.

While global warming is getting the most time and coverage today, it is only one part of the much broader issue of global climate change which is the major focus of this book.

## Chapter One
# PARADISE IN PERSPECTIVE

Is today's moderately warm, calm, and relatively moist climate typical of Earth's historical climate? Not unless you define recent as only the last 10,000 years of the 4.6 billion years of Earth's existence. We are living in an occasional and brief flicker of comfortable, nourishing, and relatively stable climate. In researching the extremes of Earth's climate history, I realized how fortunate we are to be living at this particular time.

In writing about the history of Earth's climate and the many environments that have existed over the past 4.6 billion years, *Fire, Ice and Paradise* seemed to encompass the entire spectrum of our past. For the first few hundred million years, Earth was awash in the fire of molten material that covered its surface, caused by the accumulation of the debris from space that formed our planet plus the ensuing generation of tremendous heat from the constant impacts as these smaller bodies rained down at tens of thousands of miles per hour.

As Earth cooled, the surface became hospitable enough to harbor a very primitive life, a type of bacteria that can grow in very hot water. Nearly four billion years later, evidence on all of the continents indicates that Earth had glaciers and probably sea ice over most of its surface. Thus, the period, from about 850 million to 650 million years ago frequently has been referred to as the "Snowball Earth" time. Whether the planet was ever completely covered with ice has not been unequivocally proven but this was certainly a time of a vast amount of ice. There is evidence that a significant glaciation occurred as early as around 2.2 billion years ago. So Earth had, by 600 million years ago, certainly experienced fire and ice.

Everybody has his or her own idea of "paradise", but I think you will see that, certainly for humankind, this current climate and overall environment are as good as, if not better than, at any other time in Earth's history. No one would want to lose such an environment, so let us try to understand the factors that have gotten our planet to this particular state and understand what we might be able to do or not do to try to preserve this paradise.

Fig. 4

# GRAND SUMMARY

| GENERAL CLIMATE | COMMENTS |
|---|---|
| Extreme heat, Earth's surface molten. Cooling but very hot, high $CO_2$. Still cooler by end of era. | Earth forms, then moon 50 my later. Constant ET impacts and volcanism. Oldest known rocks 4 bya. First life 3.5 bya (cyanobacteria). |
| Still warm but some glaciation about 2.3 bya. Great oxygen input at 2.4 bya, high $CO_2$. Glaciation on all continents from about 850 to 650 mya. Earth's surface almost frozen. | Impact frequency slows. Photosynthesis begins. Life creates oxygen and removes some $CO_2$. A large continent near the equator diverts currents poleward. |
| Earth warms again. Temperature higher than any time since then. Unusual time when $CO_2$ and oxygen were high simultaneously. | Dramatic decrease in ET impacts. Explosion of life in seas: crustaceans, fish-like animals and corals. One large continent below equator (Gondwana). No vegetation. Earth barren. |
| Started very warm but then began a significant cooling. $CO_2$ levels generally high and oxygen generally low. | Life rebounds. True fish, aquatic spiders, scorpions and first vertebrates. Gondwana nears South Pole. Major extinction. Possible large impact. |
| A significant glaciation occurs but was short lived. Temperature increased throughout period. $CO_2$ falls after early Silurian. Oxygen rises above today's level for first time. | First land plants and some insects appear. Earth's surface green for first time! |
| Temperature increase begins a long decline (80 my). $CO_2$ also declines sharply. Oxygen low then very long term rise (110 my). | Gondwana near South Pole. Paleo North America near equator. First amphibians. Multiple ET impacts and mass extinction end period. |
| Warm and wet early, major glaciation began 325 mya. $CO_2$ falls to very low level. Oxygen rises throughout period. | First trees. Extensive swamps early (become huge coal deposits). Extensive fires from high oxygen. Very large insects. First reptiles appear. |
| Continued glaciation. Very low $CO_2$ until end of period. Highest oxygen from Cambrian to today. Oxygen falls end of period. Temperature low early, rebounds throughout period. | One huge continent (Pangaea) by end of period. Period ends with life's greatest extinction (90%)! Mega ET impact, with crater possibly 300 miles wide. Greatest period of volcanism. |
| Very warm early. Little if any ice after 240 mya. High $CO_2$ and modest oxygen levels early, then both decline sharply. | Life recovers. Some life returns to seas after Permian extinction event. First dinosaurs, mammals and birds. Large deserts. Huge evaporate deposits (salt and anhydrites). |
| Cooler than Triassic. Modest $CO_2$ and very low oxygen levels. Lowest oxygen from Cambrian to today. | Breakup of supercontinent begins. Dinosaurs tolerate low oxygen and become giants. Another period of large deserts and extensive evaporate deposits. |
| Very warm and humid. 20° Fahrenheit warmer than today at higher latitudes. $CO_2$ begins a gradual decline and oxygen rises throughout the period. | Breakup of Pangaea continues. Dinosaurs flourish and bones are found near poles. High sea level and extensive inland seas. Huge ET impact and extinction of dinosaurs and > 65% of life as period ends. |
| Continued cooling. Glaciation at South Pole begins 35 mya and at North Pole 7 mya. Continued drop in $CO_2$ and rise in oxygen as Earth enters another icehouse stage. | Dinosaurs extinct so explosion of mammals. Man's ancestors appear. Isthmus of Panama blocks Atlantic–Pacific equatorial current routing Gulf Stream north by end of period. |
| Earth dominantly in an icehouse state. Glaciation for 90% of the time for last 800,000 to one million years. Brief interglacials every 100,000 years (like today). | Earth dynamics continue. Modern man arrives and develops the ability to reconstruct Earth's history. (Come help!) |
|  |  |

There were billions of years when humans could not have survived at all, either because of the climate or the critters. There were other periods when, had our species been present, they might have been able to survive, but it would have been in extreme discomfort. Today, the climate is such that people can survive to the ends of the Earth and comfortably thrive on most of the planet's landmasses. There are Eskimo people in the Arctic, people living in the latitudes of four seasons, those existing in perpetually warm zones of the low latitudes, and even desert dwellers.

Since modern "man" has inhabited Earth, the global climate has been colder for most of that time. Many of our ancestors of more than 10,000 years ago lived in harsh glacial conditions for 90 percent of the time. I will review other intervals of time and see how our paradise compares. Figure 4 illustrates the time in Earth's history when the temperature, $CO_2$ and oxygen levels were generally higher or lower than the levels of today. It also indicates the glacial periods, times of vast desert deposits, sea levels, approximate locations of the continents and times of large extraterrestrial impacts.

Many of us would consider a Hawaiian type of climate as the very definition of paradise. It certainly is, but as you compare climates from most of Earth's multi-billion-year history, I think you will agree that nearly all of Earth's surface today, with the exception of the two poles, basks in a relative paradise. Today's global climate provides most of us with all the things we need to nurture ourselves, to flourish, and to prosper. Sure, we can have freezing-cold days and very windy days, but the annual averages are really quite nice. A large percentage of the surface waters of the planet's oceans are at a temperature that allows us to enjoy them in comfort and relative safety, we can even submerge ourselves in the spectacular real life aquariums using scuba devices.

Starting about 12,000 years ago and going back in Earth's history for the previous 85,000 to 90,000 years, this comfort zone would have been much narrower and closer to the equator, probably south from mid-Mexico latitudes. If our ancestors had had sailing ships, the higher average wind velocities of those cold times, with the accompanying higher waves, might have been fun for the hardiest sailors but the amateurs would have found it tough going and probably stayed on land.

Going back farther in time, for nearly a million years, these 85,000 to 90,000 years of cold, windy weather persisted for all but a handful of briefer 10,000 to 15,000 -year interglacial intervals of relative warmth and calm. On land, the much drier climate accompanying these cold intervals would have caused all but a small percentage of a growing population to be underfed and undernourished.

Some say warmer climates bring on more diseases. Possibly, but tell that to the historians who have tracked the "Little Ice Age" that lasted from about 1350 A.D. to about 1850 A.D. During this 500-year period, there was plenty of disease, famine and death, accompanied by hangings of the "witches" who supposedly brought on these cold times. Huddling humanity together in cold, poorly ventilated spaces with inadequate nourishment created a disease-distributing environment. This was a time when Earth's global temperature averaged only about two degrees Celsius cooler than today!

By contrast, the "Medieval Warm Period" of "1000" years ago and the "Roman Warm Period" of "2000" years ago saw humanity flourish and increase in population and well-being. As you will see this Little Ice Age was nothing but a short interlude in the overall global warmth man has enjoyed for the last 10,000 years.

Had our predecessors existed 40 or 50 million years ago, the warmth may have been bearable and the ocean temperatures right for swimming, at least until an ancestor of *Perusarus*, a giant crocodilian creature that survived until only six to nine million years ago in (you guessed it) Peru, came along and gulped our fictitious predecessor down in one sitting if not one bite. The skull of Perusarus measures nearly six feet in length, indicating an alligator –creature probably 40 to 50 feet long and weighing 18 to 22 tons! Compare this with the Tyrannosaurus rex, which weighed in at about 7 to 9 tons. A replica of the skull of this holdover from the age of dinosaurs can be seen at the Audubon Zoo in New Orleans, Louisiana.

During the age of dinosaurs, 65 to 250 million years ago, you would not have been safe on land or in the sea. The mammals that existed were very small and spent most of the daylight hours hiding from their reptilian counterparts.

From 250 million to 325 million years ago, Earth was in the grips of another ice age and from 325 million to 430 million years ago, Earth was much warmer but oxygen content was much lower than today.

How about beyond 430 million years ago? There were no dinosaurs, but neither were there plants on the land. No greenery, whatsoever! The period when land plants finally appeared was in only the most recent 10 percent of Earth's history. Imagine a bleak, barren surface with no land-dwelling animals, not even amphibians. Nothing could inhabit a lifeless landmass upon which there was no shade or anything to eat. To put the timing in perspective, see Figure 5, a timescale highlighting key events in Earth's history. While you are reviewing Figure 5 also note the time that people have existed on Earth.

Earth was birthed in fire, sometimes dominated by ice and, I truly believe, we are now living in a paradise for our species. We have an abundance of generous grains from our grasses, nuts and shade from our trees, and on and on. How is our current paradise looking to you now?

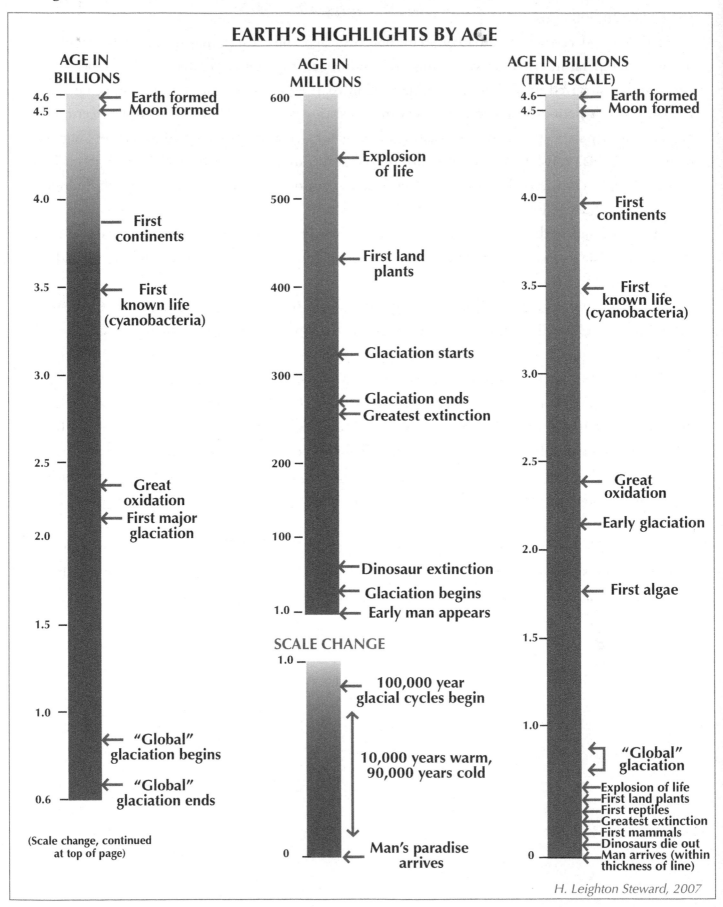

## EARTH'S HIGHLIGHTS BY AGE

**AGE IN BILLIONS**

- 4.6 — ← Earth formed
- 4.5 — ← Moon formed
- 4.0 — − First continents
- 3.5 — ← First known life (cyanobacteria)
- 3.0 —
- 2.5 — ← Great oxidation
- ← First major glaciation
- 2.0 —
- 1.5 —
- 1.0 —
- ← "Global" glaciation begins
- ← "Global" glaciation ends
- 0.6 —

(Scale change, continued at top of page)

**AGE IN MILLIONS**

- 600 —
- ← Explosion of life
- 500 —
- ← First land plants
- 400 —
- ← Glaciation starts
- 300 —
- ← Glaciation ends
- ← Greatest extinction
- 200 —
- 100 —
- ← Dinosaur extinction
- ← Glaciation begins
- 1.0 — ← Early man appears

**SCALE CHANGE**

- 1.0 —
- ← 100,000 year glacial cycles begin
- 10,000 years warm, 90,000 years cold
- 0 — ← Man's paradise arrives

**AGE IN BILLIONS (TRUE SCALE)**

- 4.6 — ← Earth formed
- 4.5 — ← Moon formed
- 4.0 — ← First continents
- 3.5 — ← First known life (cyanobacteria)
- 3.0 —
- 2.5 — ← Great oxidation
- ← Early glaciation
- 2.0 —
- ← First algae
- 1.5 —
- 1.0 — ← "Global" glaciation
- ← Explosion of life
- ← First land plants
- ← First reptiles
- ← Greatest extinction
- ← First mammals
- ← Dinosaurs die out
- 0 — ← Man arrives (within thickness of line)

*Fig. 5*

*H. Leighton Steward, 2007*

6

## Chapter Two
# THE LIBRARY IS OPEN!

Our ability to study Earth's paleoclimate history has grown rapidly. However, much of Earth's surface has not yet been researched with modern tools. What we do know has come from analyses of the rocks, sediment, ancient soils, minerals, pollen, fossils, ice cores, coral growth bands and even tree rings and leaves. These past climate indicators will be described more thoroughly in Chapter Three. The rocks and sediment that hold the information are vast in quantity and have barely been touched by the types of tools available to researchers today because so many science graduates have gone into the energy, medical or manufacturing industries. But for those who like meaningful research, scientists entering this field today will have a lifetime of discoveries coming from their probes into the past.

Since tools for what the rock, sediment, and ice records in Earth's "library" of history are telling us regarding past climates have existed for only a few decades, progress certainly will be made in our ability to translate much more from these ancient Earth records. An example of a development that allows scientists to glean more information from Earth's library was the stable isotope analysis of the mineral geothite which has helped us learn more about the carbon dioxide content of the ancient atmosphere (Professors Crayton Yapp and Neil Tabor, Southern Methodist University, personal communication). Geothite is formed in soils today as well as very old soils, and retain information about the content of the gases from the ancient atmospheres as well as its approximate temperatures. Another, relatively new, tool is found in using an isotope of beryllium ($Be^{10}$) to analyze the intensity variations of the sun and cosmic ray influences, both of which may be very important in unlocking more secrets in why Earth's climate has changed as it has.

The first geologic map that simply depicted sediment layers of the same age, was made less than 200 years ago. William Smith of England used certain fossils and observations of similar rock types to correlate from one hillside outcrop to the next. He noticed the similarities of certain fossil types in layers of similar sediment layers as he was supervising the excavation of hundreds of miles of canals that traversed the English countryside. This first map, that covered much of the near surface of England, is still surprisingly accurate. Author Simon Winchester has documented Smith's feat in his book, *The Map That Changed The World.* (Winchester, 2001) Winchester's book also demonstrates the reluctance of Smith's contemporaries to timely grasp the scientific significance of Smith's observations.

Chapter Four describes what tools are available to help researchers understand what the "books" in Earth's library contain. Hopefully, you'll never drive by a highway road cut or a natural hillside outcrop of rock without realizing how many thousands or millions of years of Earth's history are contained in those rocks and what helpful information may be residing there. Has anyone even sampled or researched that particular "page" (outcrop)? Not many of these sites worldwide have been analyzed with all the tools available today. Fortunately, several sites around the world have been extensively measured, and that is how we have obtained our current level of knowledge.

Describing how we interpret the data held in our worldwide library today is fairly straight-forward, but the integration of the various pieces of data to determine how the climate was affected is very complex. The analyses will only get better as technology continues to advance and as more of the data from Earth's library is studied.

Beyond what happened in the past, the much more difficult questions to answer are why the climate was like it was and what combination of factors (and there are many) caused the climate to change from one state to another.

Scientists have documented the technical data that exist from our very cursory peek into this library which is as large as all of Earth's outer layers, including those layers that exist at the bottom of the current and past oceans, lakes, and marshes. An expansion of the library that is accessible at Earth's surface includes the tens of thousands of cores drilled by the mining and energy industries, many from several miles deep in the Earth. With the geoscientists focusing elsewhere, few have been sampled for paleoclimate implications.

Researchers also have gone beyond earthly bounds and have measured and/or calculated the effects of the sun's varying contributions to Earth's climate, as well as the gravitational pulls and tugs on the shape of Earth's orbit around the Sun. Researchers also have uncovered direct evidence of many significant meteor and/or comet impacts that will be described in Chapter Three.

To test whether our interpretations of the causes or drivers of the paleoclimates and the changes that have occurred make good sense, some researchers make models of what may have happened by inputting the best data available. But, as mentioned earlier, where the sampling of the library has been sparse, and it generally has been, confidence in the model's output can be quite low. Thus our need to discover and analyze many more data points in Earth's library so future models will contain all the pertinent data that needs to be input to make models more reliable in forecasting past and future climates.

What does the knowledge of Earth's climate history really do for us? We hope it will allow us to use our growing array of earthly tools intelligently to try to preserve or at least extend this paradise of a climate that is now unusually beneficial to mankind's comfortable existence on Earth. Short of this goal, we hope to be able to forecast the probability and timing of, a future change so we can be prepared for the expected change.

Figure 6 gives an example of just a few of the growing list of questions that researchers have asked as they visited Earth's outcrops.

**READING EARTH'S LIBRARY**

Fig. 6

H. Leighton Steward, 2007

## Chapter Three
# CLIMATE DRIVERS

I have asked nearly everyone with whom I have conversed since I started writing this book what they think is causing Earth's climate to get warmer. The nearly unanimous answer from the non-scientist is "carbon dioxide" or "man-made carbon dioxide", except of course, from those people who are really interested in how the climate works. Read this chapter and decide for yourself. If you get discouraged trying to remember all of the drivers, just put a paper clip on Figure 7, the one-page summary, and refer back to it as needed. Yes, there are that many climate drivers!

## Sun's Luminosity or Irradiance

The sun provides nearly all of the heat that reaches Earth's surface and atmosphere. A small amount comes to the surface from heat within Earth, such as that which is emitted from volcanoes or sea-floor spreading centers. Some of the sun's heat is absorbed at the surface, some is reflected back into space, and some is radiated and retained in the atmosphere by the effect of the greenhouse gases, primarily water vapor, carbon dioxide, and methane. Besides greenhouse gases, there are many drivers that affect or modify how much of and where the sun's heat is retained. As you read on, you will understand how these other drivers come into play regarding global climate and global climate change.

As the Sun's visible energy (light) encounters Earth's atmosphere, most of this energy passes through the atmosphere and strikes the planet's surface. Clouds and dust or other particulates (aerosols) do reflect some of the visible light back into space. Of the dominant amount of the visible light that reaches Earth's surface, nearly half is absorbed, more than a quarter is reflected back into space, and the rest is radiated back as nonvisible infrared energy and temporarily "trapped" by the various greenhouses gases (Figure 8). This "trapped" heat helps keep Earth's lower atmosphere warm.

The amount of the Sun's energy that is reflected back varies as the reflectivity of the surface changes. For example, snow reflects more visible light, and dark soil reflects less visible light, so dark soils absorb more heat and then radiate more infrared heat than snow. The amount of the sun's energy that gets reflected by Earth's surface or atmosphere is referred to as Earth's *albedo*.

*Fig. 7*

# CLIMATE DRIVERS

| DRIVER | PRINCIPAL INFLUENCE | IMPACT | COMMENTS |
|---|---|---|---|
| Sun's strength (irradiance) | The amount of heat leaving the sun and headed for Earth (it varies). | Strongest | Amount retained modified by other drivers. |
| Orbital eccentricity | Determines distance from the sun at any given time. | Strong | Distance variations mean heat variation. |
| Earth's tilt | Determines Earth's seasons and heat received at high latitudes. | Strong | Additional tilt can affect polar ice melting. |
| Earth's wobble (precession) | Determines season closest to or farthest from the sun. | Strong | Can be a positive or negative feedback. |
| GHG water vapor | Strongest GHG. Affects cloudiness, albedo, vegetation, and precipitation volumes. | Strongest of GHGs | Net effect the least understood (predictable) of the GHGs. |
| GHG carbon dioxide ($CO_2$) | Captures infrared heat radiated from Earth's surface, reradiates some heat. | Strong at low saturation | Greenhouse effect non-linear. Usually follows temperature changes. |
| GHG methane $CH_4$ | Captures infrared heat radiated from Earth's surface, reradiates some heat. | Moderate (low volume in atmosphere) | Generated in wetlands, by industries and some animals. |
| Ocean currents | Distributes heat around Earth. Can change patterns quickly or very slowly. | Strong | Largest reservoir of surface heat. Affects precipitation. |
| Plate tectonics (seafloor spreading) | Causes volcanism, $CO_2$ and sulfate particle input, subuction and mountain building. | Strong long-term | Affects ocean currents, sea levels, and $CO_2$ volumes. |
| Location of continents | Affects major ocean currents and heat and moisture to poles. | Strong to weak | Land over poles promotes more glaciation. |
| Elevation of land massses | High elevations increase chemical weathering ($CO_2$ removal). | Moderate | Also affects regional climates, monsoons and locations of deserts. |
| Chemical seathering | Affects $CO_2$ removal and carbon sequestration. | Moderate long-term | Little short-term effect on climate. |
| Extraterrestrial impacts | Immediate fires, then cold for 1 to 5 years. | Strong very short-term | Can create ocean and atmospheric toxicity. |
| Volcanism | Constant source of $CO_2$, sulphate particles, and short-term soot. | Moderate to strong short-term | Provides great lava and ash layers for age dating. |
| Albedo | Determines how much solar heat is reflected or retained. | Moderate to strong | Affected by many drivers, constantly changing. |
| Fauna & flora (animal life and vegetation) | Affects albedo and oxygen, $CO_2$ and methane content of atmosphere. | Moderate | Abundance and type of flora track temperature, $CO_2$, and moisture changes. |
| Atmospheric circulation | Distributes heat and moisture and affects upper ocean current patterns. | Moderate | Winds distribute nutrients to sea life, affecting sea life abundance and carbon sequestration. |
| Cosmic rays | Suggested they create particulates that seed low level clouds (cooling). | To be determined | More research needed to verify impact magnitude. |

*H. Leighton Steward, 2007*

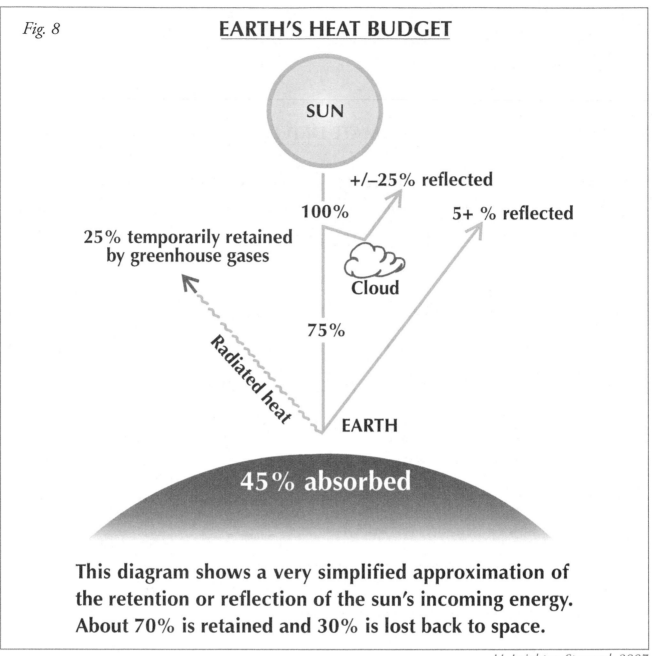

**EARTH'S HEAT BUDGET**

*Fig. 8*

SUN

100%

+/−25% reflected

5+ % reflected

Cloud

25% temporarily retained
by greenhouse gases

75%

Radiated heat

EARTH

**45% absorbed**

**This diagram shows a very simplified approximation of
the retention or reflection of the sun's incoming energy.
About 70% is retained and 30% is lost back to space.**

*H. Leighton Steward, 2007*

Earth's albedo changes just as your albedo changes if you wear a white shirt in the sunshine rather than a dark shirt. You are cooler in the high-albedo white shirt. Variations in Earth's albedo are very important in affecting global climate. Albedo will be discussed more thoroughly later in this chapter.

The energy output of the sun does vary. The best known example is the sun's cycle that recurs approximately every 11 years – the 11-year sunspot cycle. At the time of maximum sunspots on the surface of the sun, the heat released from the sun increases somewhat as a result of the presence of these dark spots that are surrounded by intense haloes (faculae) that emit an extra measure of light (heat) that is sent toward Earth. There is also a variation in how long the sunspot cycles last. To see how well the sunspot cycle length in years correlates with Earth's temperature variations, see Figure 9. The shorter the cycles, the more intense the sun's activity level.

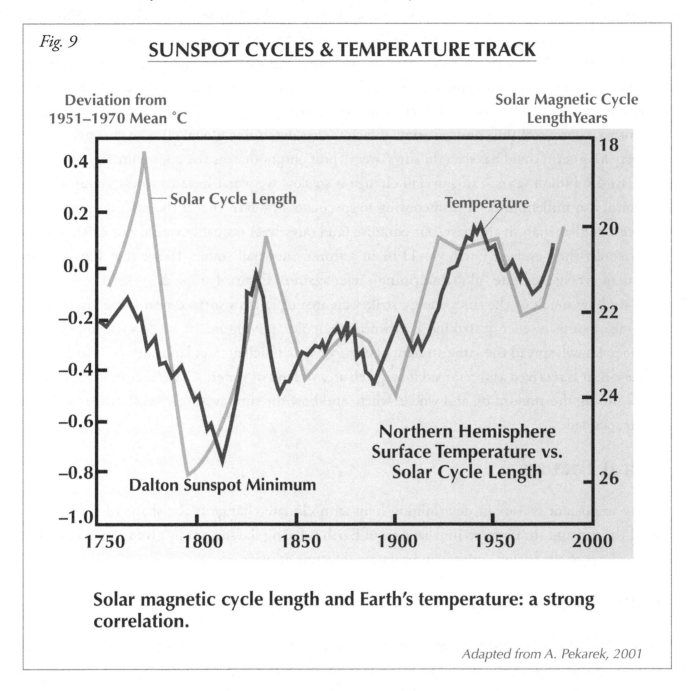

*Fig. 9*

## SUNSPOT CYCLES & TEMPERATURE TRACK

Deviation from
1951–1970 Mean °C

Solar Magnetic Cycle
LengthYears

Northern Hemisphere
Surface Temperature vs.
Solar Cycle Length

**Solar magnetic cycle length and Earth's temperature: a strong correlation.**

*Adapted from A. Pekarek, 2001*

There are other times, less well documented, when the variation in the sun's energy output is greater or lesser and it is not fully known what drives some of these longer solar cycles. It may be controlled by the changes of magnetic activity levels within the sun. We do know that solar radiation can vary beyond the common 11-year cycles. There was a 70 year long period (1645 to1715 A.D.), the Maunder Minimum, that coincided with the coldest part of the Little Ice Age, during which time sunspot activity decreased and the sun's delivery of heat to Earth diminished about one half of one percent and Earth cooled. Solar activity, represented by sunspot numbers and magnetic variations, indicates that solar activity, including outpourings of solar wind, has been unusually high for the last 70 years and the Earth has experienced an overall warming trend.

Solar intensity variation also is believed to occur in 87 and 210-year cycles (Gerhard, 2007) and the occasional coming-in phase of the cycles is believed by some scientists to cause the observed short-term cold cycles that occur about every 1,500 years. This recurring cycle, regardless of origin or duration, has been well-documented by Gerald Bond et al., 1997. That this 1,500-year cycle has been very well documented but has left so many capable scientists scratching their heads as to its cause, is just another example of the complex array of factors that determine global climate change.

A very long-term trend has seen the sun's overall heat output increase by approximately 30 percent in the last 4.6 billion years. This overall change is so slow we won't have to worry about it in the next century or millennium. It is interesting to speculate, however, that if the sun's output was not 30 percent higher than in the past, our paradise (and ourselves) probably would not exist, and our already cooler-than-average Earth would be in a frozen snowball state. To see that Earth is now cooler than average, despite "global warming", refer again to Figure 4, Page 2.

To see how much of the sun's energy strikes an area of Earth's surface near the equator versus that same amount of energy striking Earth near the poles, see Figure 10. A low angle of contact near the poles will spread the same amount of energy over a much greater land surface, which means much less heat is received and retained for a given area of land or water. Other factors are important in determining the amount of, and where, when, and how the sun's heat reaches Earth, as well how much is retained.

## Orbital Factors

Very important factors in determining long-term climate change is the shape of Earth's orbit around the sun and the changes in what part of Earth is facing the sun at any given time. Following are the effects of the various orbital and rotational factors at play:

## Orbital Eccentricity

If Earth's orbit around the sun was perfectly circular, which it sometimes nearly is, Earth would receive the same amount of solar heat each year for its entire 365-day orbit around the sun. If, however, the orbital path was more elliptical, Earth sometimes would be closer to the sun during part of the year and sometimes farther away, and thus would not receive a constant level of solar heating. This is exactly what happens. Earth's orbit gradually changes from nearly circular to somewhat elliptical and back over a period of approximately 100,000 years (Figure 11). Earth is currently in a more circular-shaped orbit, although it is always somewhat elliptical.

Today the variation in Earth's maximum and minimum distance from the sun in its 365-day orbit is as much as 3 million miles. While that may not seem like much in an average distance of 93 million miles, it amounts to a modest difference in heat received and apparently has been strong

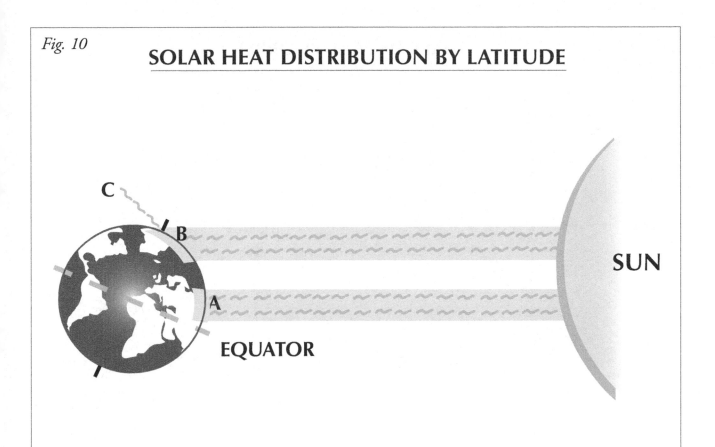

*Fig. 10*

## SOLAR HEAT DISTRIBUTION BY LATITUDE

**For a given medium, say water, more of the sun's heat is absorbed per square meter near the equator (A) than near the poles (B) because an equal amount of the sun's energy is spread out over a greater surface area (in effect, diluted) near the poles. Also, the lower angle of contact with the surface results in still more of the sun's rays being reflected back into space (C).**

*H. Leighton Steward, 2007*

enough in its extremes that, when combined with what part of Earth receives heat and when it is received (i.e. summer or winter), it has caused the last eight or ten glacial - interglacial cycles that have occurred over these 100,000 year intervals. Studies of ice cores reveal that the orbital factors have been the primary drivers of Earth's major climate changes for nearly the last several million years and undoubtedly for eons past to a greater or lesser extent.

To observe the astonishing regularity of the last four glacial – interglacial cycles, which have been very well documented through the studies of ice cores from Antarctica, see Figure 12. All but the last 10,000 years, the current warm period, occurred without any influence by the human race.

As Paul Mayewski and Frank White pointed out in their book *The Ice Chronicles*, (Mayewski, P.A. and White, F., 2002) the more order we see in nature, the better chance we have of predicting

*Fig. 11*

# SHAPE OF EARTH'S ORBIT

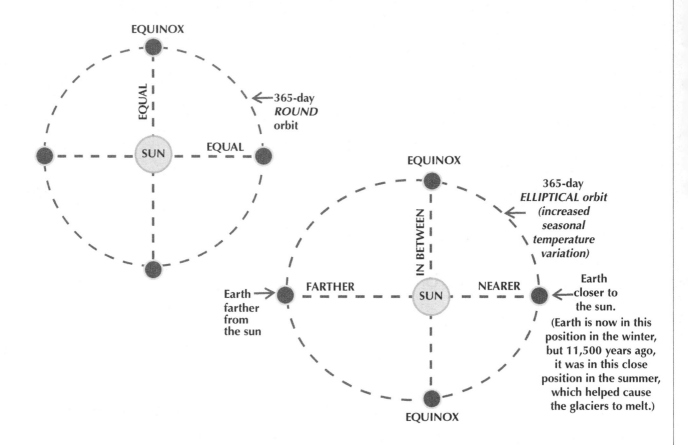

There is approximately 100,000 years between Earth's most round and most elliptical orbit around the sun. This is about the same amount of time as there is between the glacial and interglacial climates on Earth for at least the last 800,000 years. The difference in the closest orbital position of Earth to the sun today and in Earth's farthest position from the sun is about 3 million miles, which affects the amount of energy that strikes Earth's atmosphere.

*H. Leighton Steward, 2007*

the future. I like their observation. However, I also know that some wise soul said, "Nothing ever stays the same". One day, the influence of a combination of climate drivers will converge and take us out of this recent rhythmic cycle and onto a new one. For our future generations, let's hope it morphs into an extended period of paradise! Unfortunately, with the ever-changing moving parts such as orbital variations and solar intensity changes, I doubt the permanent paradise will ever occur unless mankind achieves powers beyond my current imagination. But then, look at what man has thought would never occur but has happened - lights, flight, atomic power, moon walks, landing on asteroids, a tiny injection to prevent horrible diseases, and talking to someone anywhere on Earth or in space from a tiny handheld device.  High-tech is in its infancy; some of us can remember no air conditioning or television!  What's next?  Anti gravity?  Warp speed?  Time travel?  Why preclude controlling the climate?  But let's get back to today.

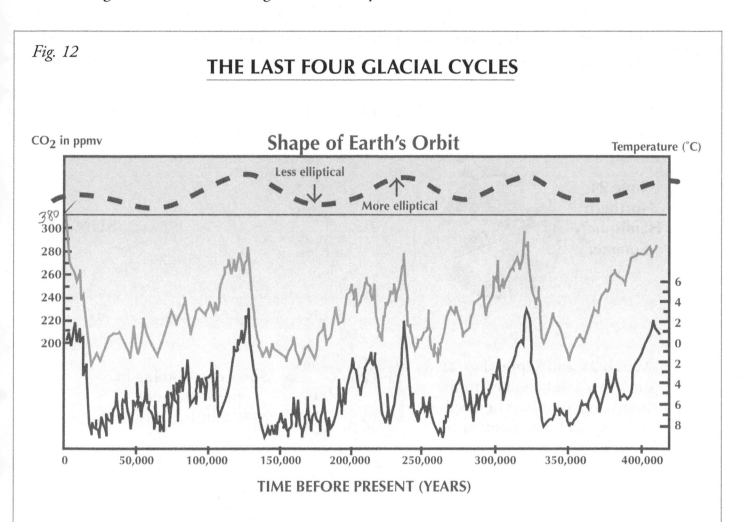

*Fig. 12*

# THE LAST FOUR GLACIAL CYCLES

**Antarctic temperature and $CO_2$ history from ice core analyses. Note the amazing rhythmical similarities of the four cycles, which indicate the very strong solar-orbital influences on Earth's climate. Since Earth's $CO_2$ levels do not drive the solar-orbital cycles, you can see why many scientists doubt the currently popular "$CO_2$ CAUSES global warming" argument. On the other hand, an increase in $CO_2$ does provide some positive feedback (support) to a warming Earth, but the magnitude of that effect is still being debated.**

*Adapted from Pettit, et al., 1999*

# Earth's Tilt Axis

Earth's axis of rotation is aligned with the north and south poles and is actually tilted relative to Earth's orbital plane around the sun (Figure 13). This tilt is what gives us the seasons. If the equator was lined up directly with the plane of our 365-day orbital path around the sun, the sun's rays would fall on exactly the same latitude with the same intensity for the entire year and there would be no winter, spring, summer, or fall. Many astrophysicists think that a Mars-sized body that collided with Earth and formed the moon changed or tilted Earth's axis of rotation and gave Earth's its seasons.

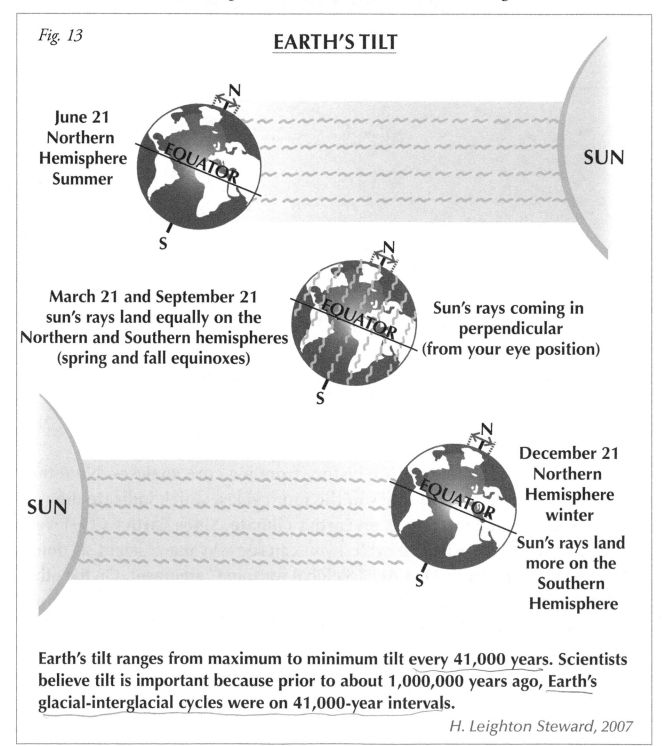

*Fig. 13*

## EARTH'S TILT

**June 21 Northern Hemisphere Summer**

**SUN**

**March 21 and September 21 sun's rays land equally on the Northern and Southern hemispheres (spring and fall equinoxes)**

**Sun's rays coming in perpendicular (from your eye position)**

**SUN**

**December 21 Northern Hemisphere winter**

**Sun's rays land more on the Southern Hemisphere**

**Earth's tilt ranges from maximum to minimum tilt every 41,000 years. Scientists believe tilt is important because prior to about 1,000,000 years ago, Earth's glacial-interglacial cycles were on 41,000-year intervals.**

*H. Leighton Steward, 2007*

The tilt angle is currently about 23.5 degrees, but it varies over time, from 21.8 to 24.4 degrees, in about a 41,000-year period. This change in tilt does have an effect on the climate by increasing or decreasing how much of the sun's energy lands on Earth's poles. For instance, the greater the angle of tilt, the more of the sun's energy shines down on the poles, which, in the summer, can contribute to additional melting of ice on the pole tilted toward the sun at that time. Obviously, the opposite pole would be a little colder in its winter, because more of that pole's ice or land would be hidden from the sun's rays. If other drivers are collectively exerting a warming or cooling influence on Earth, the tilt angle could be more important in helping to cause or in dampening the magnitude of a climate shift.

An example of the possible major influence of changes in Earth's angle of tilt lies in the fact that beyond the last one million years, Earth's glacial periods occurred at approximately 41,000 year intervals; the same timing as one complete tilt cycle.

## Earth's Wobble

Like the slowing of a spinning top, Earth also wobbles with one full wobble every 19,000 to 23,000 years. This wobble causes the seasons to change depending on whether the time when the sun is closest to, or farthest from, Earth falls in summer or in winter. (Figure 14). Again, this can have an effect of increasing or decreasing what spot on Earth receives the most or least heat and in what season. Since there is more low-albedo (less reflective) water in the Southern Hemisphere and more highly reflective landmass in the Northern Hemisphere, when and where the sun is delivering its heat can exert a significant influence on heat absorption and global climate.

These variations in the length and intensity of heating of different parts of Earth's geography clearly have strong effects on the climate. For instance, 11,500 years ago, the time when Earth was the closest to the sun in Earth's eccentric orbit occurred in July, which caused very hot and lengthy summers and, combined with orbital and tilt influences, resulted in the melting of most of Earth's ice cover from the last ice age. While the winters were very cold, the ice and snow could not accumulate fast enough to keep up with the intense summer melting (again, see Figure 11, page 16) and Earth was temporarily freed from the grips of that ice age and transformed into the current period of paradise. Astronomer's ability to reconstruct ancient orbital positions mathematically has helped us to understand these global climate changing relationships.

*Fig. 14*

# EARTH'S WOBBLE (AXIAL PRECESSION)

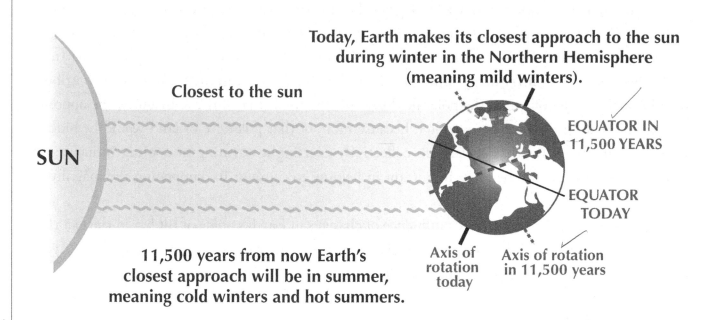

Today, Earth makes its closest approach to the sun during winter in the Northern Hemisphere (meaning mild winters).

Closest to the sun

SUN

EQUATOR IN 11,500 YEARS

EQUATOR TODAY

Axis of rotation today

Axis of rotation in 11,500 years

11,500 years from now Earth's closest approach will be in summer, meaning cold winters and hot summers.

As Earth wobbles, the season when Earth is closest to, or farthest from, the sun in Earth's elliptical orbit changes (precesses). Today, Earth is closest to the sun in the winter. In 11,500 years, Earth will be closest to the sun in the summer. Since the average albedo (reflectivity) is different in the two hemispheres (more mid-latitude land in the Northern Hemisphere), this also can have an impact on how much heat is absorbed by Earth for the winter (or summer) season. Will everything be exactly the same in a full cycle (23,000 years)? No, because the orbital eccentricity (100,000-year cycle) and Earth's tilt (41,000-year cycle) will be in a different alignment. Wobble is just another of the ever-changing orbital orientations that helps determine Earth's constantly changing climate.

*H. Leighton Steward, 2007*

## Orbital Summary

Any of the orbital, tilt or wobble variations can have an effect on Earth's climate. Since they all complete one full cycle of change or influence over differing periods of time, 100,000, 41,000 or 19,000 to 23,000 years, you can easily imagine that at any particular time, they may reinforce each other or, in a different alignment, tend to cancel each other out.

To have all of the orbital factors maxed out toward a colder effect or maxed out toward a warmer effect would be akin to having a "perfect orbital storm" influence on climate at that particular time and push it very strongly in the direction of a colder or warmer climate. Fortunately, since astronomers can calculate when the various alignments will occur that will favor warming or cooling, this will help our distant descendants prepare for such changes.

For the geological near term, hundreds to a few thousand years, the effect of the orbital factors and resulting variations in solar intensity would suggest an influence toward a somewhat cooler Earth than our paradise of the last 10,000 to 12,000 years. This is another good reason to sequester (store), in a retrievable place, the greenhouse gases we create from the use of fossil fuels. If Earth begins to cool, we will be able to reach for our stored blanket of $CO_2$, however modest its effect, to help us stay a little warmer or to increase food production from the critically important greenhouses.

## Greenhouse Gases

Hooray for greenhouse gases! Greenhouse gases are very important in retaining the heat radiated from Earth's surface. Unlike other gases present in Earth's atmosphere, such as nitrogen (78 percent) and oxygen (21 percent), which trap almost none of the sun's heat, the greenhouse gases trap nonvisible, infrared heat that is radiated back into the atmosphere from Earth's surface and thus helps warm the atmosphere. Without them, Earth would be frozen, since most of the sun's incoming energy simply would be reflected and radiated back into space. We don't, however, want too much effect from the greenhouse gas content in the atmosphere, just as we don't want too little of it. Mankind's paradise for the last 10,000 years has had just about the right amount of the combination of these gases to help keep the near-surface temperature at an average of 59 degrees Fahrenheit.

Humans have some tolerance for variations in the principal greenhouse gases, possibly much more than we realize. For instance, humans can live in good health at sea level, where the atmospheric carbon dioxide content averages around 0.038 percent, and methane about 0.00002 percent and the oxygen level (not a greenhouse gas) averages around 21.0 percent. Humans can even survive at 15,000 feet where the respective levels are about the same but the total amount is less because of the overall reduction of all gases caused by the altitude and thinness of the atmosphere. We are a lot less tolerant of changes in oxygen and methane ($CH_4$) than changes in water vapor or $CO_2$. The two other greenhouse gases, nitrous oxide ($N_2O$) and ozone ($O_3$) have much less effect on Earth's climate, even at sea level, because they are present in such small amounts.

## Water Vapor ($H_2O$)

Water vapor is by far the most dominant of the greenhouse gases, accounting for as much as 95 percent of them. (Figure 15). It occurs in forms and places that can contribute to either heating or cooling on a global or regional scale. High, bright cloud layers made up of ice crystals temporarily "trap" some heat and thus contribute to high-altitude warming, whereas dense, low-level clouds reflect some of the sun's incoming ultraviolet heat. This has a cooling effect by reflecting sunlight from the low cloud's bright tops. These low-level clouds can also temporarily "trap" some of Earth's radiated, infrared, heat. This ability to trap and reradiate heat is most noticeable on cloudy nights, when the temperature stays much warmer than on cloudless nights, when Earth's radiated heat can escape to space. When the clouds build up to tremendous heights, some heat energy is lost to space and results in cooling. The answer to the question *"Does water vapor cool or heat the Earth?"* is *"Well, it depends."* Atmospheric water vapor and cloud formation are other areas ripe for research, because they can have large temperature effects, and their net effect is still not well understood.

The average amount of water vapor in Earth's atmosphere today ranges from near zero to four percent. The general content of water vapor in the atmosphere can be inferred by a well-known law of physics: warm air can hold more water vapor than can cold air. Thus, the cold episodes are generally drier (and windier), and the warm periods are generally more moist (with lower average wind velocities). Analyses of ice cores and worldwide sediment cores that cover multiple cold and warm cycles have confirmed this generalization. I know of no scientist who disagrees because the ice-core data that show huge amounts of dust being deposited during the cold periods are very convincing.

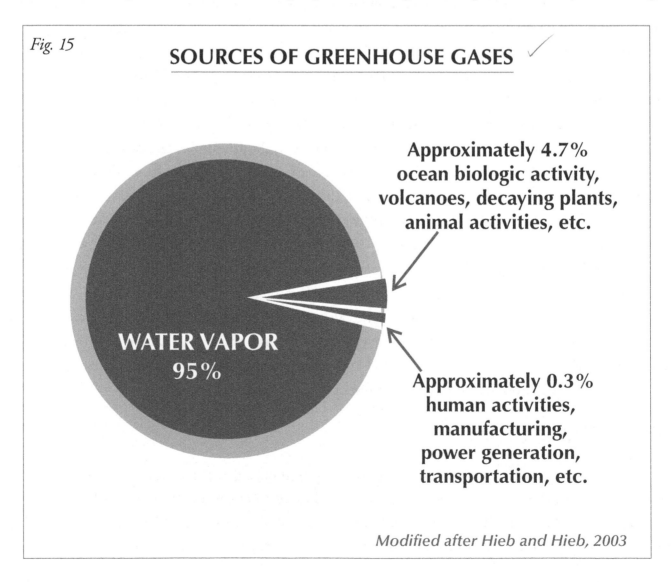

*Fig. 15*

## SOURCES OF GREENHOUSE GASES

Approximately 4.7%
ocean biologic activity,
volcanoes, decaying plants,
animal activities, etc.

**WATER VAPOR
95%**

Approximately 0.3%
human activities,
manufacturing,
power generation,
transportation, etc.

*Modified after Hieb and Hieb, 2003*

*Fig. 16*

# THE WARMING EFFECT OF
# ATMOSPHERIC CARBON DIOXIDE

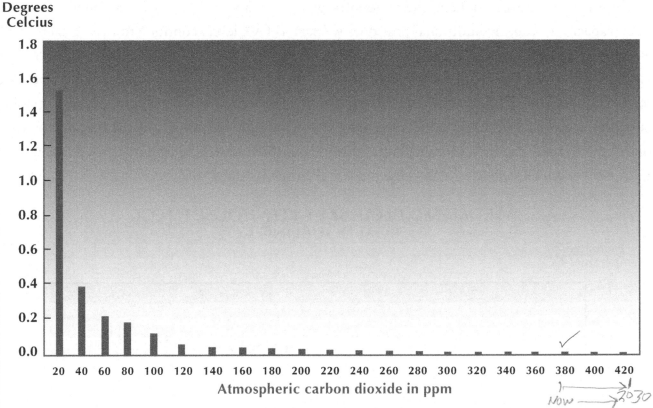

**Degrees Celcius** (y-axis)

Atmospheric carbon dioxide in ppm (x-axis)

*Now* → *2030*

Anthropogenic warming is real; it is also minuscule. Using the MODTRANS facility maintained by the University of Chicago, the relationship between atmospheric carbon dioxide content and increase in average global atmospheric temperature is shown in this graph.

The effect of carbon dioxide on temperature is logarithmic and thus climate sensitivity decreases with increasing concentration. The first 20 ppm of carbon dioxide has a greater temperature effect than the next 400 ppm. The rate of annual increase in atmospheric carbon dioxide over the last 30 years has averaged 1.7 ppm.

From the current level of 380 ppm, it is projected to rise to 420 ppm by 2030. The projected 40 ppm increase reduces emission from the stratosphere to space from 279.6 watts/m2 to 279.2 watts/m2.

Using the temperature response demonstrated by Idso (1998) of 0.1°C per watt/m2, this difference of 0.4 watts/m2 equates to an increase in atmospheric temperature of 0.04°C (<0.1°C). Increasing the carbon dioxide content by a further 200 ppm to 620 ppm, projected by 2150, results in a further 0.16°C increase in atmospheric temperature.

*1800*

Since the beginning of the Industrial Revolution, increased atmospheric carbon dioxide has increased the temperature of the atmosphere by 0.1°.

*I. David Archibald, 2007*

# Carbon Dioxide (CO$_2$)

Carbon dioxide temporarily absorbs heat in Earth's atmosphere by capturing some of the nonvisible, infrared energy that is reflected back toward space. Simplistically, one would think that the more CO$_2$ in the atmosphere, the more infrared heat is captured in the atmosphere. However, CO$_2$'s capacity to capture additional heat does weaken as CO$_2$ levels continue to rise. Exactly how much more or less heat is captured and retained by various levels of CO$_2$ and whether CO$_2$ is a principal driver of global climate or is simply a greenhouse gas that may provide a modest positive feedback (support) for heat retention is at the heart of the debate currently being waged about global warming. After considering the detailed analyses of multiple ice cores and calculations of CO$_2$ logarithmic (rapid) decline in its capability to trap additional heat, I believe that if CO$_2$ is having an impact on the climate it is likely quite modest (Figures 16 and 17).

**Fig. 17**

## ESTIMATED CLEAR SKY GREENHOUSE EFFECT FROM DOUBLING CO2

Estimated Temperature (°C)

Charnock & Shine (+1.46° C)
Kondratjew & Moskalenko (+0.87°C)
Lindzen (+0.64°C)

PPMV CO2

**These projections support Archibald's hypothesis on major CO$^2$ effect occurring at very low levels of CO$^2$.**

*Modified after S. Milloy, 2007*

However, even if $CO_2$ at its present atmospheric concentration is not contributing significantly to the current warming trend but is only a weak positive feedback factor, $CO_2$ is still an important greenhouse gas that, particularly at low concentration in the atmosphere, provides a portion of the overall greenhouse effect that helps keep Earth from being a frozen planet.

Before someone tries to get you to think of $CO_2$ as a pollutant, also consider that it is absolutely critical to supplying the entire spectrum of plants with their required energy to proliferate and start the food chain for nearly all forms of life on Earth.

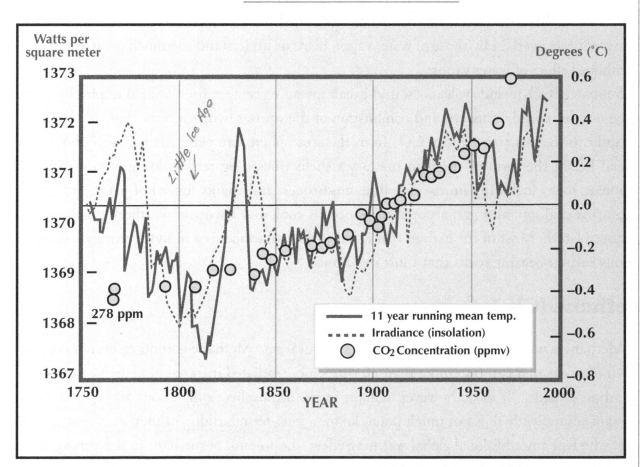

**Fig. 18**

## NORTHERN HEMISPHERE TEMPERATURE VS. SOLAR IRRADIANCE

Legend:
— 11 year running mean temp.
···· Irradiance (insolation)
○ $CO_2$ Concentration (ppmv)

**CO2 and temperature don't co-vary. Solar energy and temperature do co-vary.**

*Adapted from Hoyt, 2006*

The $CO_2$ level has varied dramatically in the last 500 million years, ranging from the geologically recent glacial period lows of approximately 180 ppm, to the preindustrial level of 280 ppm (today we are at 380 ppm), to a time around 500 million years ago when the level was approximately 7,000 ppm, or 18 times higher than today. Global warming skeptics cite the fact that glaciations have occurred during periods with high $CO_2$ levels as evidence that $CO_2$ levels were certainly not the drivers of the climate changes and that the predicted logarithmic decline of $CO_2$'s ability to trap heat is correct. This latter observation allows us to understand more easily how glaciations and high levels of $CO_2$ can coexist. It also explains the reason temperature trends and $CO_2$ trends do not move in lock-step. Again see Figure 4, page 2, for an illustration of the generalized $CO_2$ and temperature variations over time. While there is a gross correlation with temperature changes over the long-term, detailed studies over somewhat shorter time periods show that the relationship often breaks down (Figure 18). The fact that there is so much more water vapor in the lower atmosphere than $CO_2$ (Figure 15, page 22) and that the principal infrared heat absorption bands of $CO_2$ are almost entirely overlain by those of water vapor, helps us understand the much greater heat-trapping greenhouse effect of water vapor.

Sources of $CO_2$ include volcanoes, mid-ocean spreading centers, fires, animal respiration, warming of the oceans, cosmic impacts, and combustion of the various hydrocarbons that we use as fuel.

Sinks, or factors that remove $CO_2$ from the atmosphere, are plant life through photosynthesis, animal life in the form of mega to micro sea shells (including reef-building organisms), calcium carbonate rocks including limestones, lime mudstones, and chalks, burial of plant and animal life that forms coal, oil, and gas, absorption as oceans cool, and chemical weathering processes (to be discussed later). Most of the carbon that is not in the atmosphere or in living things is stored in the various carbon-bearing rocks that I just mentioned.

# Methane ($CH_4$)

Methane is what is commonly referred to as natural gas. Methane is another, even more powerful greenhouse gas that can trap more than 20 times more radiated infrared heat than an equal volume of carbon dioxide. It does, however, remain in the atmosphere only about 10 years and absent a constant resupply, there is not much potential for a long-term buildup of methane. Fortunately, for those who fear any additional global warming effect, the amount of methane in the atmosphere today is 200 times lower by volume than current $CO_2$ levels.

Methane levels increase as swampy areas on Earth increase, because plant life rotting in water creates methane. Also, it has been reported recently that living plants may emit relatively large volumes of methane; other researchers dispute these claims. However, if substantiated, isn't it amazing that this fact wasn't known and documented until 2006? See what I mean about new discoveries awaiting our future researchers? As Earth warms and the air becomes more moist and aids additional precipitation, additional swampy areas should develop. Extensive rice farming over the last several thousand years has caused a measurable rise in methane levels, as have the emissions of livestock. Termites also emit significant quantities of methane.

The tundra region releases methane when the environment turns warmer and the previously frozen methane deposits are released to the atmosphere. Methane is stored in the cold tundra peat bogs as methane hydrates or clathrates (simply frozen crystals of methane). Frozen methane is also found at specific temperature and pressure environments in sediments at or just below the sea floor. Subsea methane volumes can be huge and their sudden release can have a warming influence, albeit temporary, on global climate. These deposits can be released by submarine landslides, submarine volcanism, or even by very large impacts from space that directly destabilize the temperature and pressure environment that is harboring the methane.

# Ozone (O$_3$)

Ozone is a trace gas present in very small amounts (parts per billion) in Earth's atmosphere. Ozone's most important role relative to human life is that it absorbs incoming ultraviolet light and also protects us from skin cancer and damage to our eyes.

Its lesser role, beyond absorbing the ultraviolet rays, is that it does cause a small amount of warming at the near surface but causes some cooling in the stratosphere and thus deserves mentioning. But for our conversation about factors that can have a significant impact on global climate change it will not be discussed further.

# Nitrous Oxide (N$_2$O)

Nitrous oxide is also a trace gas present in very small quantities (parts per billion), and while it captures some radiated heat coming from Earth's surface, its contribution appears to be very small and likely does not drive global climate change.

# Greenhouse Commentary

Now that we know what the three chief greenhouse gases are and what they do, you may wonder why all the fuss over a little variation in carbon dioxide content, especially considering that some well-respected scientists like H. M. Fischer et al., 1999, U. Siegenthaler et al., 2005, N. Callion et al., 2003, Monnin, et al., 2005 and Muddlesee et al., 2001, have determined that the changes in CO$_2$ levels during recent glacial cycles FOLLOWED, not CAUSED, the temperature changes and the lag in the CO$_2$ response was several hundred years as determined by all of the researchers. An example of the temperature and CO$_2$ relationship is demonstrated in Figure 19. By the way, ditto for methane increases FOLLOWING the warming and cooling. I believe that prior to Fischer, et al.'s discovery of CO$_2$ lagging the temperature changes, so many scientists and public figures had been convinced of the dominant role of CO$_2$ as a climate driver and had so convinced the general public, as well as myself, that there has been a reluctance on the part of the large majority to admit that the $CO_2$ role is much less than previously thought.

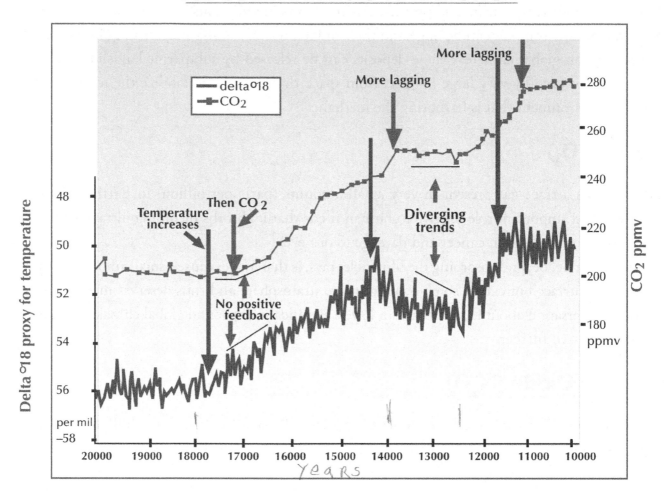

*Fig. 19*

## ICE CORE ANALYSES–ANTARCTICA

**This detailed analysis clearly shows the lag of $CO_2$'s response to a prior temperature change. A cause does not <u>follow</u> an effect. It also illustrates that $CO_2$ and temperature do not always co-vary (see 14,400 to 12,500). There is no indication of positive feedback from around 18,200 to 14,200.**

*Concordia (Dome C) Analyses after L.E.Monnin, et al., 2005*

Another climate scientist, Dr. Sherwood B. Idso, (2001), points out that the warmer than "normal" *Medieval Warm Period* and the cooler than "normal" *Little Ice Age* both occurred in the last 1,000 years accompanied by no significant change in $CO_2$ levels. Idso also performed several real-world observations over a 20 year period while he was working at the United States Water Conservation Laboratory in Phoenix, Arizona. The results of this research indicated that the climate of Earth is not very sensitive to further increases in $CO_2$ and his best estimate of doubling $CO_2$ in the atmosphere from 300 to 600 ppm would probably be only about 0.4 of a degree Celsius.

The crux of today's concern is that no one wants the temperature to change much in either direction. We know greenhouse gases temporarily trap some heat and, if increased enough, can cause or help the atmosphere to warm by re-radiating some of that captured heat. A reduction in greenhouse gases can allow Earth to lose some heat to space and help cause cooling unless other climate drivers, as many scientists believe, overpower the greenhouse effect. A significant change in either direction and our climate could move out of the paradise zone. While some scientists are still struggling with whether increases of $CO_2$ beyond the current levels have significant warming effect, let's not continue to add $CO_2$ at such an unprecedented rate, because we still do not know the precise answers as to exactly how much it may cause some other effect such as making the oceans more acidic. Increased acidity could possibly affect the development of the shells of some of the life in the ocean that are at the beginning of Earth's food chain.

While it is probable that some warming will occur with the input of more of the greenhouse gases, we don't exactly know what magnitude of change it will take to push the temperature out of the paradise zone. Many people have decided that any man-made additions are bad and have endorsed dramatic decreases in greenhouse gas emissions. While I can understand this view, I endorse a more phased reduction in greenhouse gas emissions at a rate that will not cause a significant reduction in mankind's standard of living and the amount of money we spend on climate research, medical research, improving appliance efficiencies, developing alternate energy sources that make sense and the like. Currently, however, there is no technology that can sequester $CO_2$ at a reasonable price or in a secure reservoir from which the extremely soluble $CO_2$ cannot escape. This is another area where more research dollars should be provided because if the technology becomes available to sequester $CO_2$, a storehouse of $CO_2$ could prove useful when Earth begins to cool.

Earth's orbital orientation is now slowly changing from a configuration that promoted recent interglacial warming to one that will favor global cooling. Many scientists, including Professor William Ruddiman at the University of Virginia, (Ruddiman, William F. 2001) have suggested that the only thing that may have kept us out of the next glacial cycle, has been man's agricultural success over the last 8,000 to 10,000 years, which has generated additional methane (a stronger greenhouse gas than $CO_2$) and the release of $CO_2$ through the burning of fossil fuels (coal, oil, and natural gas).

So, while most people are trying to minimize the current greenhouse gas buildup, let's boost our climate-change research so we will know more about how and when to use any means at our disposal to temper the sudden and significant climate changes such as those that have occurred over and over and over again in Earth's natural, prehuman history.

Of interest is a non-greenhouse gas, oxygen. In paleohistory, long, long ago, oxygen ($O_2$) levels were at one time zero. Earth first became oxygenated significantly about 2.4 billion years ago. Oxygen levels later rose, about 275 million years ago, to make up approximately 30 percent of the atmosphere and fell at another time, about 200 million years ago, to approximately 12 percent of the atmosphere. When oxygen levels were very high, wildfires abounded and when the levels got very low, many species went extinct. Oxygen levels today are about 21 percent, which is ideal for us and the other current inhabitants of the planet. When considering the role of $CO_2$ in feeding the plants, remember that plants provide us with the oxygen we breathe.

# Ocean Currents

Approximately 70 percent of Earth's surface is covered with water. On the average, water absorbs much more and reflects much less of the sun's incoming heat than does land. The oceans hold a tremendous amount of heat and are very important in affecting Earth's climate.

The oceans give off heat when the water is warmer than the overlying atmosphere which is why many coastal areas have more temperate climates than land located farther from the coast. Conversely, if the water is cooler than the atmosphere, it will tend to cool the otherwise hot atmosphere and also temper the adjacent coastal climates.

With effects from varying wind patterns, gravitational effects on the tides, differences in water salinity, and differences in temperature, the oceans develop currents that transport these warmer or cooler waters around the globe. A significant change in these current patterns can greatly influence Earth's climate in a regional, if not global, manner.

What lets us predict when the world's major ocean currents will change patterns? At this moment, the answer is much more than I or anyone can tell you, because this remains an understudied, under-measured area of science. One obvious influence is the changing locations of the continents, but ocean currents change even when the continents have hardly moved. The data gathering is increasing rapidly, and this is another area ripe for discovery.

Besides simply affecting the temperature, ocean currents can affect the amount of moisture in the air. Warmer air can hold more moisture than colder air. So, if you want more rain or snow, having more moisture in the air is very helpful.

Since colder air doesn't hold as much moisture as warm air, if a warm current should flow up into the northern latitudes and provide a little more moist air, you should expect to get more rain or snowfall than normal. This is what may have happened when the warm Gulf Stream began to flow to the North Atlantic when the opening to the Pacific ocean closed in Panama three million years ago. In somewhat cooler times, this moist air that overlies the warm Gulf Stream has allowed the glaciers and ice sheets to grow more rapidly and amplify the ice buildups during the glacial cycles.

In contrast, cold-water currents upwelling adjacent to a mid-latitude shoreline can create a colder, low-water-vapor atmosphere and cause very dry conditions that result in little rain or snow. This is what is happening just onshore in the Atacama Desert (the driest place on Earth) on the west coast of South America, even though the land area is directly adjacent to the Pacific Ocean.

Over time, usually a long interval such as tens of thousands or millions of years, the major ocean currents do change their pattern of flow, and in some instances, these changes cause major climate changes. Again, let's look at the Gulf Stream as an example of how a change in current flow or position can change or help change the climate. As pointed out by Wally Broecker of the Lamont-Doherty Earth Observatory at Columbia University (Broecker, W. S., 1991 and 1992) the Gulf Stream has not always flowed to the far north as it does today. When it has ceased to have a warm surface flow as far north, that sector of the world - Greenland, Iceland, northern Europe - has generally gotten colder. It has gotten enough colder that despite there being less water vapor present, what little snow does fall will not tend to melt in the orbitally-controlled shorter summers and glaciers slowly will begin to build.

While changes in the patterns of major ocean currents may take many thousands to millions of years, this shift from a far-north-flowing Gulf Stream to a flow ending farther to the south appears, from the detailed studies of ice cores in Greenland, to have occurred in a much shorter time. This change may have been caused by large volumes of lighter, salt-free fresh water supplied by melting glaciers or ice sheets 8,000 years ago to make the saltier, heavier Gulf Stream sink below this lighter fresh water. (See Figure 20 to observe the magnitude of the possible change in location of the northern extent of the Gulf Stream.)

So what if it gets a littler colder in Greenland and northern Europe? The problem is that there is currently no massive amount of artificial heat available to heat the fields and help grow crops in a much colder climate. This is an example of a large regional climate change. Don't forget that the interplay of many other factors (orbital, albedo, location of continents, greenhouse gas content, etc.) can act or act together to cause the onset of a world wide or regional cooling or warming. If the world's climate cools enough to affect the world's agricultural output significantly, the wars and chaos caused by trying to squeeze more than six billion people into a small band around the equator will make any previous world war insignificant in comparison because everyone on Earth will be fighting for what little productive land remains.

There are very short-term changes in some surface current patterns that modify Earth's regional climates. The shift from a central Pacific warmer El Niño surface current to a cooler La Niña surface current is an example of changes that can and do occur every few years. Even these relatively small changes in ocean surface temperature cause changes in the climate from North and South America to Africa and either help provide moisture or wreak havoc by causing drought in certain areas of those regions.

*Fig. 20*

# CREATION OF THE GULF STREAM THREE mya

Five mya, an open passageway existed between North and South America. Since this very warm water was not circulating northward, the north pole was very cold and dry, and glaciers grew very slowly because of the lack of moisture for snow production.

By three mya, tectonic forces caused land to be uplifted and close the passageway between North and South America, and the westward flow of warm water got diverted northward, where it delivered moisture necessary for large amounts of winter snow and rapid glacial expansion during all but the interglacial, warm cycles like the one we are living in today.

## PLATE TECTONICS

The next several drivers are related to plate tectonics that cause a variety of horizontal and vertical continental movements that affect Earth's climate.

# Location of the Continents

The location of the continents can make a big difference in the world's climate. As we have just seen, continent positions can block the paths or create new paths for the ocean currents. (See Figure 4, page 2, for a generalized summary of the location of the continents and equatorial ocean currents at various times over the last two billion years.)

Landmasses located near the poles can hold large volumes of glacial or sheet ice. During the most recent ice age, ice covered not only the north pole but also vast areas of land over most of Canada and portions of what is now the northern United States and northern Europe. Many scientists have estimated that ice was approximately 1.8 miles thick just north of what is now New York. The estimates are made by measuring the amount and rate of rebound (uplift) of Earth's surface from where it had been depressed by the weight of the overlying glacial ice.

Continents located directly over a pole can provide a cold solid base upon which to build huge volumes of ice. Once land at the pole gets cold and covered with ice, it remains frozen until a major warming occurs. When water instead of land is present at a pole, the ocean circulation constantly delivers some heat to the base of any ice that has formed, and the ice will not tend to build up nearly as thick as on land. Since more than 85 percent of the ice that floats in the ocean is beneath the ocean's surface, like the iceberg that sunk the *Titanic*, the elevation of the ice will never be very high. This lack of elevation means that temperatures over a landless pole will be higher than where ice has built up to higher (colder) elevations on landmasses that overlie a pole. Temperature everywhere on Earth declines with increased altitude. As a current example, the continent of Antarctica at the south pole holds more than seven times as much ice volume as that present at the open ocean of the north pole, even when including the ice accumulated on the landmass of nearby Greenland.

When the landmasses covering part of the surface of Earth are broken into many scattered pieces, a relatively uninterrupted flow of ocean currents can occur. This allows heat to be distributed in latitudinal bands, with warm surface currents found at lower latitudes and cooler surface currents found nearer the poles. When these perpetually cooler waters deliver little moisture to the overlying atmosphere, glaciations and sheet ice grow very slowly. If the summers are relatively hot and melt the ice each year, no build-up will occur, unless land's altitude results in permanent freezing.

Alternatively, when the landmasses coalesce into one or two supercontinents or a large landmass is located directly over the equator, warm equatorial currents can be blocked and diverted poleward, which takes more heat and moisture in that direction. This moisture is then available to feed the larger snowfalls and thus ice buildup.

The supercontinents also tend to harbor vast deserts in their interiors. These vast deserts that replaced earlier, plant-covered, smaller continents have a much higher reflectivity coefficient (albedo) than a plant-covered surface. While the deserts themselves are hot, I believe the net effect of more of the sun's energy being reflected back into space should result in a somewhat cooler average global temperature.

# Continental Drift or Sea-floor Spreading

What causes the location of the continents to shift? It has been called continental drift, plate tectonics, or sea-floor spreading. Below the continents, Earth's crust is approximately 20 miles thick, whereas below the sea-floors, the crust averages only about 3.5 miles thick. The continental plates, several of which carry continental landmasses with them, "drift" or move around on a molten subsurface as a result of heat convection cells that rotate within this molten layer (forces caused by heat flow within Earth), as well as possible effects from Earth's rotation or gravitational influences by the sun, the moon or the other planets. (Again, look at the Grand Summary chart, Figure 4, page 2, to see how the continental locations have changed over the eons.)

The rate of the plates' movement is only an inch or two or three a year, but over millions of years, these distances can become thousands of miles. The separation of the Americas from Europe and Africa, which began approximately 175 million years ago, has been spreading at a relatively low rate compared with other plate movements but now has produced the Atlantic Ocean which is up to 4,500 miles wide between South Africa and lower South America.

Of course, the plates not only spread but also collide. Where these spreading rates and the accompanying collision rates are very high, Earth produces inordinately large earthquakes, tsunamis, volcanic activity, and mountain building. Even at a modest collision rate, the Himalayas are an example of exceptionally high mountain building that is being caused by the northward drift of India into the continental Asian plate. This collision began approximately 55 million years ago and continues today. The collision has caused the uplift of the 15,000-foot-high Tibetan plateau, which, along with the Himalayas, has a very strong effect on the moisture distribution over a large portion of that region. Another active area, one with a more rapid collision rate, lies under the ocean west of Indonesia where we frequently witness large earthquakes, tsunamis, and volcanic activity. You might remember reading of the huge volcanic explosion of Krakatoa in 1883, often referred to as "the shot heard around the world", and, of course, the very large Indonesian earthquake of 2005 that caused a tsunami that killed more than 200,000 people in the same general part of the world.

Where these truly huge masses of land (continental) or sea-floor (oceanic) plates collide, one plate will override, and one will underride (get subducted) beneath the other (Figure 21). The overriding plate usually gets squeezed up into mountain ranges such as those along the North and South American west coasts, where those continental plates are overriding the oceanic crustal plates of the eastern Pacific.

Where one plate under-rides another, that crust is pushed down, reheated to a molten state, some of which is squeezed back up through the newly formed volcanoes as lava with its associated gases (again, see Figure 21). So, the whole plate-tectonic scene either can create new land or mountains, or it can destroy existing crust of Earth by recycling it back down into the molten layer of Earth with some of it being extruded back up through volcanic activity. Relating this phenomenon back to one of Earth's greenhouse gases, much of Earth's carbon dioxide is released into the atmosphere by this relatively constant volcanic activity, a factor we cannot control, and also by the degassing of the oceans when Earth's temperature has risen. Today, some of the increase in $CO_2$ is occurring as a result of human-related emissions while the rise in Earth's temperature also is probably contributing to the increase of atmospheric $CO_2$.

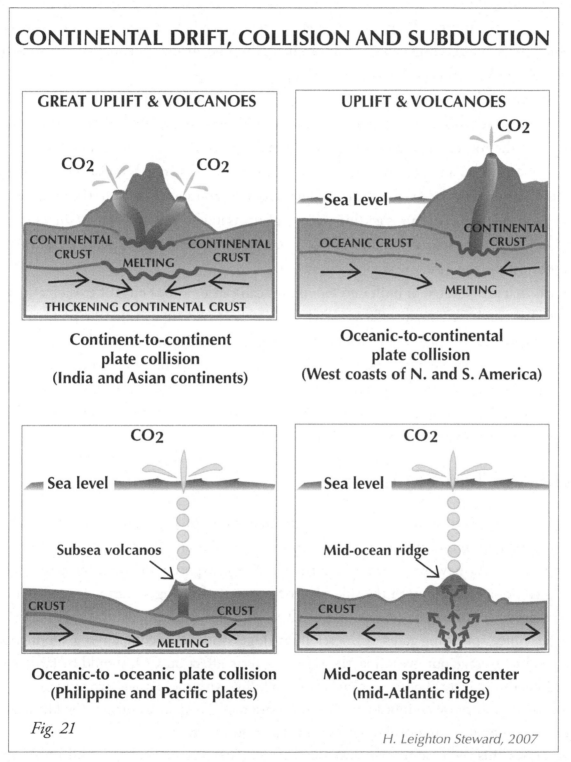

# CONTINENTAL DRIFT, COLLISION AND SUBDUCTION

**GREAT UPLIFT & VOLCANOES**

**Continent-to-continent plate collision (India and Asian continents)**

**UPLIFT & VOLCANOES**

**Oceanic-to-continental plate collision (West coasts of N. and S. America)**

**Oceanic-to-oceanic plate collision (Philippine and Pacific plates)**

**Mid-ocean spreading center (mid-Atlantic ridge)**

*Fig. 21*

*H. Leighton Steward, 2007*

At the mid-ocean spreading centers, when spreading rates are high, the rising molten rock causes the sea-floor to swell along these multi-thousand-mile long ridges. This significant elevation of the sea-floor can displace water onto the continents and alter (lower) Earth's reflectivity factor or albedo. Alternatively, with low spreading rates and a much lesser swell at the spreading centers, sea level will tend to be lower and Earth's albedo higher as a result of more land surface being exposed. Also, $CO_2$ coming from the spreading centers will be more modest and result in a lower greenhouse effect.

*What does all of this have to do with climate?* **A lot!** We have seen that continental locations and size can affect ocean currents, location and extent of deserts, and plant life. Also, continents provide bases upon which to build huge volumes of ice, including the effects of increased elevation and the resulting cooling to promote still more ice buildup.

Elevation also affects atmospheric circulation and rain or mountain snow and glaciation. Most of us know that when, a wind blowing from west to east encounters a mountain range, the upper west flank and the top of the range will get more moisture in the form of rain or snow because of the cooling effect on warm air rising. Since cold air cannot hold as much moisture as warm air, the moisture is released. On the east or downwind flank of the mountains, most of the moisture already has been precipitated, plus the air is getting warmer again as it descends and can retain what moisture is left and not be precipitated as rain or snow. As long as the mountain ranges exist, the leeward side of the ranges will be the drier areas.

The areas that are elevated or built into mountains can cause various rates of erosion and chemical weathering that can have climate effects, which will be discussed later in this chapter.

Ultimately, the active plate movements and continental collisions that build or replace eroded land are the only reasons we have land upon which to live in this paradisial climate we are currently experiencing. Otherwise, the mountains would eventually all erode away, and we would have a shallow layer of water covering the entire Earth! Never thought you would look favorably at land building by earthquakes and volcanoes did you?

## Elevation of Land Masses

The height to which land is elevated - high mountains or high plateaus - not only affects atmospheric circulation and moisture distribution but also has an effect on physical and chemical weathering rates. As new rock or sediment surfaces are exposed to the atmosphere, chemical weathering causes a reaction with the siliceous rocks (quartz-bearing sandstones or granitic rocks) and, in the process, removes large volumes of $CO_2$ from the atmosphere, and that helps cool the atmosphere. Since spreading rates of the sea floor ridges and the accompanying subduction zone trenches, tend to average out over time, the very long-term effect on $CO_2$ should be fairly constant. However, sometimes, instead of the oceanic plates colliding with each other, two or more continental plates will collide and cause an inordinate rise in elevation of a large area, such as the Himalayas, and greatly increase the rate of removal of $CO_2$ from the atmosphere.

What causes this increase in chemical weathering? Steep slopes collapse and break up easily, earthquakes are quite prevalent in the mountain-building phases that last for tens of millions of years, water gets between the cracks or fractures in the rock and freezes and mechanically breaks up the rock, and glaciation grinds up large volumes of rock into much smaller fragments and exposes many more square centimeters of fresh surface to be chemically altered.

About 55 million years ago, the plate bearing the continent of India began to collide with the continental Asian plate. The collision is still occurring. The collision continues to build an unusually high plateau that is being actively eroded, the Tibetan plateau. The collision also has built the tallest perpetually cold mountains on Earth, the Himalayas, which are still slowly growing in elevation. Look at Figure 22 and observe how much the temperature and average $CO_2$ level have declined since this uplifting began. Other ideas for this long-term decline in temperature and $CO_2$ include the carbon being stored in new "limestone" rocks, lower sea-floor spreading rates of the Tertiary period having pumped less $CO_2$ into the atmosphere and also causing sea level to be lower, resulting in a higher reflectivity of the sun's energy back into space, or that the temperature decline is a cosmic ray effect.

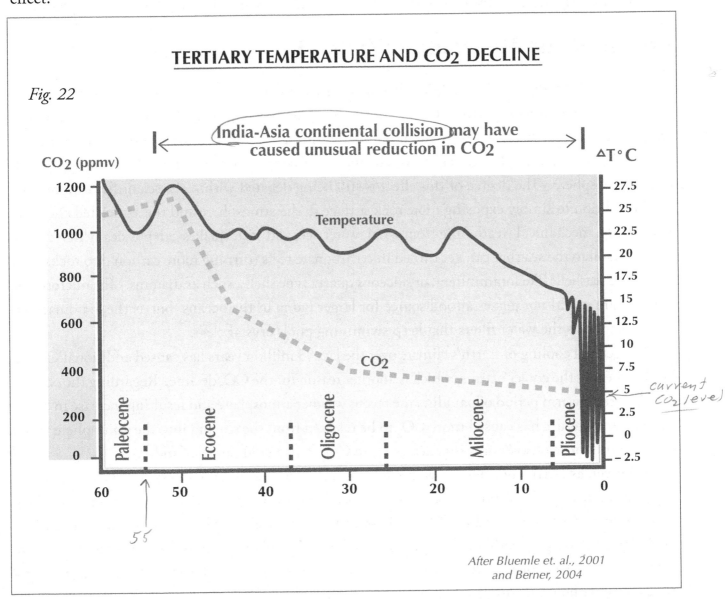

**TERTIARY TEMPERATURE AND CO2 DECLINE**

*Fig. 22*

After Bluemle et. al., 2001 and Berner, 2004

No major collision had occurred for the previous 185 million years, and Earth's temperature and $CO_2$ content became very high in this quiet interval. Spreading rates and the resulting sea level were also higher 100 million years ago. Most scientists believe that little permanent ice is known to have existed beyond the poles during this mid-Cretaceous warmth. Farther back, between 325 and 240 million years ago, many continent-to-continent collisions occurred which allowed elevation of landmasses and much more chemical weathering and long term $CO_2$ removal. Also, with the supercontinent named Pangaea blocking equatorial current circulation which helps with additional precipitation of snow, Earth cooled and evidence shows that ice covered the poles and extended to latitudes below 40 degrees. This makes an argument for having some long-term effect by $CO_2$ and therefore some possible impact on global climate unless the reduction in $CO_2$ was simply caused by the substantial drop in temperature during the ice age of that time, just as $CO_2$ levels have been shown to follow the temperature declines of the last several glacial cycles. The latter seems more plausible.

## Chemical and Mechanical Weathering

Many paleoclimatologists believe that mechanical weathering rates, which affect chemical weathering, can have a long-term influence on Earth's climate. The principal reason is that mechanical weathering, such as that caused by elevation of land through plate collision, exposes a lot of new or fresh rock debris that can be acted upon by chemical weathering and lower the carbon dioxide content of the atmosphere. The degree of this effect is still being debated within the scientific community.

In addition to simply exposing more rock surface to the atmosphere and the associated chemical weathering, mechanical weathering from wind, water, ice, and earthquakes can physically move more rock and soil to the sea that can get turned into carbonate reefs (turning more carbon into rocks) and carbonate seashells like foraminifera, or siliceous quartz type shells, such as diatoms. The microscopic diatoms are useful not just as a food source for larger fauna in the oceans, but as the diatomaceous Earth that coats the water filters that keep swimming pools crystal clear.

The general cooling of Earth's climate over the last 55 million years has caused additional $CO_2$ to be absorbed by the cooler oceans, which is another reason for the $CO_2$ decline. Regarding the oceans, in our current warm period of paradise, the recent warmer atmosphere and resulting increase in ocean surface temperature has caused more $CO_2$ to be released from the sea and into the atmosphere much like a warm bottle of soda quickly releases more $CO_2$ than a cold bottle of soda.

However, as Earth continues to warm and the ice sheets and mountain glaciers melt and expose long-buried bedrock, the process of chemical weathering on these surfaces can resume and help slowly remove $CO_2$ from the atmosphere. Less $CO_2$ results in slower plant growth which means less of the greenhouse gas methane being created and also, less plant cover allows for a more reflective land surface and a cooler Earth. These are examples of other negative feedbacks that help prevent a runaway greenhouse warming.

# Extraterrestrial Impacts

Look at the surface of the moon to get some feel for the number of times Earth has been hit by objects from space. With the Earth having a much larger cross section in space and also having a much larger gravitational attraction on approaching space objects, Earth has been hit many more times than the moon. We can catalogue many impact craters on the surface of the land or the sea-floor but they are just a minuscule number compared with the total that has penetrated our atmosphere (Figure 23). Why don't we see all of their scars? Erosion gradually wears away their rims and then the craters themselves or fills their centers and then covers everything with a new layer of soil or rock. Also erasing the scars is the effect of plate tectonics that continually destroys (subducts) old oceanic crust. With 70 percent of Earth's surface covered with water, most of these tsunami-creating impacts occurred in the oceans. Meanwhile, new crater-free oceanic crust is being created at the mid-ocean spreading centers by the out-pouring of new molten rock. For an excellent data source of where nearly 200 of the impact craters are located on Earth's surface, refer to the University of New Brunswick (2008) source, referenced in the Bibliography.

*Fig. 23*

## EXTRATERRESTRIAL IMPACT CRATERS

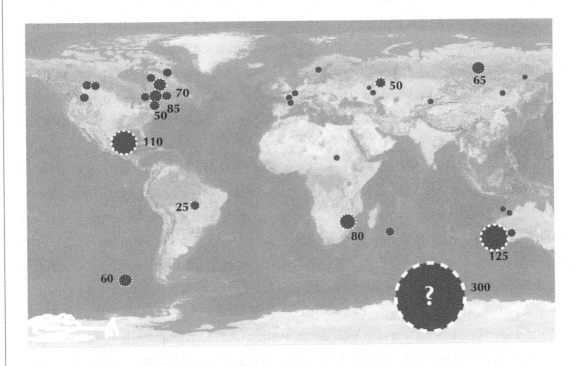

**Known impact craters more than 15 miles in diameter with mileage shown for all more than 25 miles across. More than four times this number of craters have been identified with craters less than 15 miles in diameter. Scale is relative only crater to crater.**

*Personal communication with D. Abbott, 2008
and R. Von Frese, 2006*

Most objects that enter Earth's atmosphere are traveling between 40,000 and 60,000 miles per hour (11 to 16 miles per second!). Most rifle bullets only travel at 2,000 to 3,000 <u>feet</u> per second. If just an auto-sized boulder struck your neighborhood at 60,000 miles per hour it would wipe out not only you and your neighborhood but likely your entire city.

The most recent space rock of super size (about six miles across) struck Earth in just such a manner 65 million years ago and apparently helped kill off the dinosaurs. The crater it left beneath the Yucatan Peninsula in Mexico has been measured at approximately 110 miles across. It has been speculated that the wave it caused hundreds of miles away, still had to be more than 1,000 feet high as it crossed above Florida to account for the size of the boulders the water carried into the western Atlantic Ocean.

The impact and resulting heat wave, wild-fires and mega-tsunamis undoubtedly killed millions of dinosaurs but the major extinction event was probably the <u>sudden climate change</u> caused by the dust blocking the incoming energy from the sun, causing a "nuclear" <u>winter that persisted for several years.</u> Plants died from cold and/or a lack of sunshine. Also, there would have been a large amount of acid rain triggered by the impact. Volcanic activity was high at this time, probably caused or aided by, the cataclysmic impact. All of these deadly effects caused by the impact contributed to the <u>extinction of more than 60 percent of the species on Earth.</u>

While many would say this impact-driven climate change was temporary, its destruction of much of the world's continental fauna and flora, as well as a lot of sea life that previously interacted with the carbon content of the atmosphere, must have had some lingering effect on climate. Since $CO_2$ levels have fallen gradually since the impact, the fauna and flora obviously rebounded within a reasonably short geological period of time, and the flora, through photosynthesis, helped remove some of the $CO_2$ that was released when the <u>110-mile hole was blasted into the thick sequence of carbonate rock ($CaCO_3$).</u>

Even large, older impact sites have been found on Earth. All of <u>the large ones, except for those before life appeared on Earth,</u> seem to be coincident with <u>periods of climate or atmospheric changes and the extinction of many species.</u> While the long-term climate change may not seem to be as significant as that brought about by some of the key long-term climate drivers, the short to intermediate effects can be very dramatic. Given some positive feedback from other drivers, impacts may have helped move the world into a new climate state.

New impact sites are continually being discovered. Researchers from Ohio State University, including Professor Ralph Von Frese, (Von Free, Ralph, 2006) using high-resolution gravity data, have found what appears to meet some of the criteria for an impact crater. Located about a mile beneath the sea-floor and under an ice sheet near the coastline of eastern Antarctica (Wilkes Land), the anomaly measures approximately 300 miles wide. And if it does prove to be an impact crater, the object from space that created it is calculated by Professor Von Frese to have been about 30 miles wide! The location of this anomaly is shown with a question mark in Figure 23. Should the estimated date of about 250 million years ago be correct, it is probably the <u>smoking gun</u> of the event that ended the Permian period and resulted in the greatest extinction of life ever recorded on Earth. It would have had an impact on global climate, both immediately and, through eliminating plant and sea life, for thousands of years to follow.

Another possible impact crater, about 125 miles wide, that occured about 250 million years ago, has been located off the continent of Australia. Did a huge comet or asteroid break up and hit Earth with a double whammy? More research probably will tell us.

A much smaller and more recent impact crater "only" 18 miles wide has been found in the Indian Ocean in 12,500 feet of water about 900 miles southeast of Madagascar. Huge, unusual, wedge-shaped deposits call chevrons are found onshore that "point" toward the reported impact site. These instantly deposited wedges of sediment stand 600 feet high, are found three miles inland, and contain fossils of deep-water origin that are fused with metals typical of impact sites. The tsunami is estimated to have been more than 600 feet high to account for the elevation of the chevrons. Compare this with the 2005 tsunami in Indonesia, which was less than 100 feet high.

Dr. Dallas Abbott, a research scientist at Lamont-Doherty Earth Observatory at Columbia University, and a leader in the research on this recent crater, reported in the *New York Times* on November 14, 2006, that the time of impact was thought to be about 4,800 years ago (Abbott, Dallas, 2006). While Earth did experience a drop in temperatures that lasted several hundred years about 4,500 years ago, there is no current evidence that the two events are related, but time will tell. More evidence of its effect on Earth surely will be found. To put the impact in historical perspective, this was just before the great pyramids of Egypt were built. That something from space struck Earth that recently and created a hole 18 miles wide beneath 12,500 feet of water further legitimizes researchers' attempts to figure out how we might deal with the next such object headed Earth's way.

Dr. Abbott is also a champion of the idea that the huge extraterrestrial impacts likely caused periods of enormous volcanic eruptions, a hypothesis that I also prefer unless new data prove it incorrect. Dr. Abbott says the statistical analysis by the group at Lamont has determined, with 97 percent confidence, that nine of ten of the largest volcanic fields that have been identified in Earth's history have been coincident with a large extraterrestrial impact. (See Figure 23 again, page 39, for the location of identified impact craters more than 15 miles in diameter.). There have been more than 125 impacts identified on Earth with craters less than 15 miles in width.

# Volcanism

Volcanic eruptions also can affect Earth's climate and the extent of the effect is generally proportional to the magnitude of the eruption, the duration of the eruption, and the type of the eruption (explosive versus simply flowing). Examples range from the relatively gently flowing Mauna Loa volcano in Hawaii, to a somewhat explosive Mount Saint Helens in the northwest United States, to Krakatoa in Indonesia, which in 1883 blew the entire top off a mountain and caused Earth to go without summers for a couple of years. Volcanically ejected glass from Krakatoa is present in the growth bands of star corals located in the Florida Keys. Then there was the Toba volcanic explosion, about 74,000 years ago in Indonesia, which blew 200 cubic miles of debris into the air and almost wiped out our ancestors. Some geneticists have speculated that only about 10,000 people survived the Toba event and nearly all of them were located in one region in Africa. That certainly would make us all not-to-distant cousins who have chosen to have serious family fights rather than just getting along with each other. Evidence elsewhere indicates other isolated pockets survived.

Two hundred and fifty million years ago, a vast volcanic field in Siberia, the Siberian Traps, covered an area the size of Alaska. This volcanism, which was approximately coincident with another giant impact(s), changed the makeup of the atmosphere and probably caused or helped cause the extinction of most species on Earth. One more example of volcanism would be the mega-eruption at what is now Yellowstone Park which occurred about 630,000 years ago. That supervolcano exploded over a relatively short period of time, dumped ash in the Pacific Ocean and the Gulf of Mexico, and spread a layer of ash a foot thick 750 miles away in Iowa. (Smith R. and Siegel, L. J., 2000)

These larger blasts or long lasting volcanic eruptions cause the climate to change in interesting ways. The initial blast of dust, debris, and small sulphate particles immediately cools Earth by blocking (reflecting) the sun's rays from reaching Earth's surface. A few years later, however, other climate drivers take control again and help cause Earth to continue warming or cooling. Long-term vast but nonviolent eruptions that continuously put more $CO_2$ and sulphate particles into the atmosphere obviously affect the climate in opposite directions and the net effect is not yet well known. What exactly happened, and when and why, are in the Earth's "library" just waiting to be discovered by some paleoclimate researcher.

## Albedo

Albedo is a measured percent of the reflectivity of Earth's surface at any given time or sun angle. Clean snow or ice or pure white clouds have a very high albedo of around 90 percent for direct overhead sunlight, while the lowest albedo of around five percent that absorbs most of the sun's energy occurs in dark forests, dark soils, and sunlight striking the oceans or large bodies of water from directly overhead (Figure 24).

As the snow gets dirty, the clouds get darker, and light colored soil gets covered with grass, plants, or water, the albedo will drop. While it is easy to imagine snow or ice having a high albedo, direct overhead sunlight striking the desert or a white sandy beach also can have a high albedo (as high as 50 percent).

Albedo is a very important component because it determines how much of the sun's energy is absorbed at Earth's surface and is thus very important in affecting Earth's climate. Cloud cover, the amount of water versus land, and the amount of plant cover versus desert surface that the sunshine strikes from a high angle are probably as significant to the total Earth albedo as are the volumes of very high-latitude ice and snow cover.  An exception to this would have occurred when the glacial periods experienced ice cover all the way down to the mid-latitudes. To visualize the albedo-changing effect of a past high-sea-level stand that inundated many continents where low-albedo water covered previously higher-albedo land, see Figure 25, page 44.

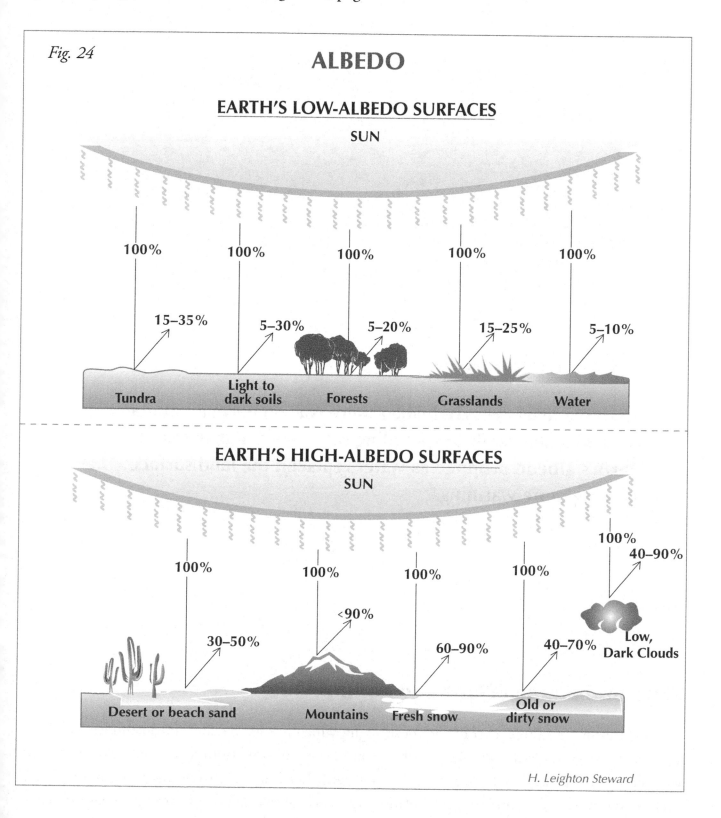

*Fig. 24*

# ALBEDO

## EARTH'S LOW-ALBEDO SURFACES

SUN

100%   100%   100%   100%   100%

15–35%   5–30%   5–20%   15–25%   5–10%

Tundra   Light to dark soils   Forests   Grasslands   Water

## EARTH'S HIGH-ALBEDO SURFACES

SUN

100%   100%   100%   100%   100%

30–50%   <90%   60–90%   40–70%   40–90%

Desert or beach sand   Mountains   Fresh snow   Old or dirty snow   Low, Dark Clouds

*H. Leighton Steward*

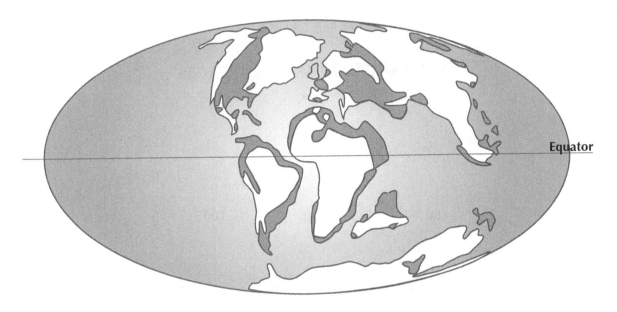

## SEA LEVEL 100 MILLION YEARS AGO

Equator

◼ INUNDATED AREAS

**High sea level occurred when there was no permanent glacial ice and when seafloor spreading rates were higher than today. Earth's albedo dropped as water replaced the land surface, causing more warming.**

*Adapted from D. Merritts et al., 1998*

*Fig. 25*

# Flora and Fauna (Plant and Animal Life)

Grass, shrubs, and trees absorb a lot of sunlight, which results in a low albedo and warmer land surface. These same plants take the greenhouse gas carbon dioxide from the air and release oxygen. This then tends to help cool the atmosphere and keep Earth at a relatively moderate temperature level. As stated earlier, an abundance of plants and trees that live and die in swampy areas generates a lot of methane which is also a greenhouse gas that will tend to capture some of the infrared heat radiated from Earth's surface.

I have heard people say that their plants grow better if they talk to them every day. While this old saying is thought by many to be an "old wives' tale", it underscores the benefit of $CO_2$ to plants, because we release $CO_2$ with every breath we exhale. Animal life, including people, takes in oxygen and gives back $CO_2$.

To digress a moment, the Supreme Court of the United States recently ruled that the Environmental Protection Agency (EPA) has the authority to rule that $CO_2$ is a "pollutant" and can be regulated. One could ask why the same rule does not apply for oxygen which can be toxic to pilots who breathe 100 percent oxygen for extended periods of time. People breathe oxygen and plants "breathe" $CO_2$. If the people who believe "$CO_2$ is bad" are successful in causing Earth's people to try to reduce severely the amount of $CO_2$ in the atmosphere, we will not have to worry about robust plants in our houses and agriculture in our fields, nor an overpopulated Earth because there will not be enough to eat! With no scientists on the Supreme Court, it is little wonder that the judges come up with non-sensical views based on "evidence" provided by a very politically motivated Intergovernmental Panel on Climate Change (IPCC). The country in the world that stands to lose the most in economic strength by following a set of rules voted in by foreign officials is the United States. Do you doubt that they would mind seeing the United States reduced in world power and influence? The IPCC, at the highest levels, is made up predominantly of social scientists and political appointees.

Sea life is also important in Earth's carbon cycle, because many of the forms of plankton and algae absorb carbon from the waters to make their calcareous shells. As the life forms die, the shells sink to the bottom and effectively sequester (store) the carbon there for long periods of time. This can lower atmospheric $CO_2$ and tend to have a cooling effect.

Meanwhile, volcanic activity associated with subduction zones or the sea-floor spreading centers does release new $CO_2$ into the atmosphere. Over geologic time, more $CO_2$ has been sequestered in rocks than has been put back into the atmosphere. The "proof" of this lies in the fact that $CO_2$ levels four or five hundred million years ago were several thousand parts per million versus only 380 parts per million today.

## Atmospheric Circulation

Just as the ocean currents distribute heat and some overlying moisture around the surface of Earth, the winds also affect where heat and moisture are distributed. Winds blowing across the oceans such as El Niño and La Niña have regional if not global effect on the climate for periods of several years. Winds can drive surface currents that can create upwelling of deeper waters that provide the nutrient-rich environments for the growth of smaller sea life. This microscopic life removes $CO_2$ from the atmosphere by incorporating the carbon in their shells. These nutrient-rich zones also support an abundance of fish that feed on these smaller animals.

Winds blowing across continents play a very important role in distributing moisture to areas far removed from the coastlines. Since atmospheric moisture content affects cloud formation and precipitation, winds can affect temperature and vegetation patterns. While wind patterns are usually an important factor in regional climates, these regional climates all add up to affect global climate. Wind enhanced rainfall also causes more chemical weathering, which, along with vegetation, can remove $CO_2$ from the atmosphere. For a very simplified illustration of Earth's major atmospheric circulation patterns, see Figure 26.

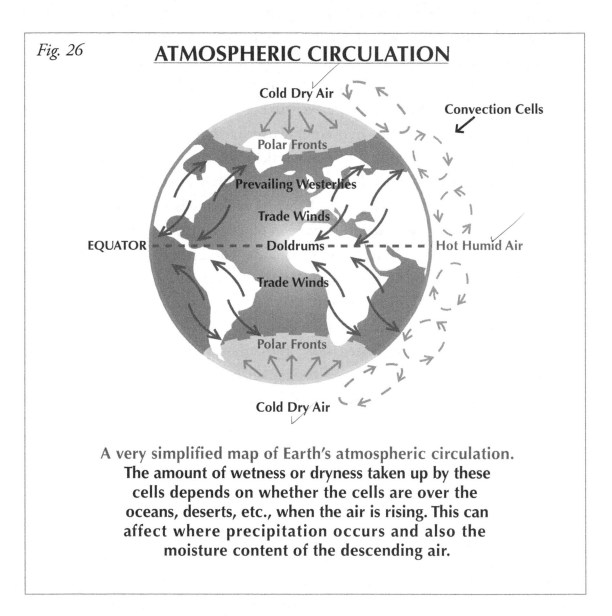

*Fig. 26*

## ATMOSPHERIC CIRCULATION

A very simplified map of Earth's atmospheric circulation. The amount of wetness or dryness taken up by these cells depends on whether the cells are over the oceans, deserts, etc., when the air is rising. This can affect where precipitation occurs and also the moisture content of the descending air.

## Earth's Rotation

The rotation of Earth is not often listed as a climate driver, but I think it deserves mentioning. An Earth with no spin would mean a half-fried, half-frozen Earth. The half that received sunlight gradually would change at the rate of 1/365th of the surface each day as Earth made its annual orbit of the sun but the result would be essentially the same; one half would be temporarily fried, while the other side would be temporarily frozen.

The spin of Earth affects the atmospheric circulation patterns as well as the oceanic circulation patterns, which we know are both distributors of the heat that affects Earth's climate.

The rotation is "*constant*" (in quotes only because Earth's rotation is gradually slowing down). After the collision that gave birth to the moon, Earth's rotation was faster. A day on Earth lasted only about six hours. Don't try to use this bit of trivia in calculating tomorrow's climate change; it isn't worth it! The 16 hours of slowing has taken 4.6 billion years in which to occur.

# Cosmic Rays

I first heard of cosmic rays when I was very young. Comic-strip and Saturday matinee characters used cosmic ray guns as they traversed the universe. While all this was very intriguing, I had no idea that cosmic rays really existed. Well, cosmic rays really do exist and it has been proposed that they are important in determining when many of Earth's global climate changes occur. Rather than being shot from fictitious cosmic ray guns, these truly atomic bullets originate and get sprayed through space when suns, much more massive than our own, ultimately collapse and explode with unimaginable force (a supernova). With trillions of such suns in the universe, supernova explosions occur very frequently. Most of the explosions happen far, far away and have no discernible effect on Earth but some blow up "nearby" and shower our solar system with cosmic rays.

What we are interested in is what effect, if any, these cosmic rays have on Earth's climate. Dr. Henrik Svensmark, (Svensmark, H. and Calder, N., 2007), director of the Center for Sun-Climate Research at the Danish National Space Center, and other scientists such as Jan Veiser of the University of Ottawa and Nir Shaviv of the Hebrew University of Jerusalem (Shaviv, N. J. and Veizen, J., 2003), believe the answer is that cosmic rays do have a significant effect on the climate by controlling the amount of low-level cloud cover on Earth. Svensmark has backed up his beliefs with some fairly crude but nevertheless encouraging experiments that seem to support the hypothesis that cosmic rays that survive the trip through the sun's and Earth's magnetic shielding and Earth's upper atmosphere, cause an ionization as the rays, traveling near the speed of light, collide with molecules in the lower atmosphere. These newly created, very small particles then provide the nuclei on which water droplets can form, which then can create low-level clouds. The low-level cumulus clouds have highly reflective tops that can and do reflect much of the sun's energy back into space and, in doing so, help cool Earth. The greater the volume of cosmic rays that reach Earth's lower atmosphere, the more collisions would occur and create more small particles that seed cloud formation and the more clouds would form to provide this cooling. Conversely, fewer cosmic rays reaching Earth's lower atmosphere would mean fewer clouds and more sunlight reaching Earth's surface, which would then cause Earth to warm.

Most cosmic rays that enter the solar system do not make it to Earth's surface. They are repelled predominantly by the magnetic field of the sun through its solar wind. They are also repelled to a lesser extent by Earth's own magnetic field. Still other cosmic rays are eliminated as they collide with molecules in Earth's upper atmosphere, where the clouds that form there do not have the same cooling effect, because the high clouds are mainly made of ice crystals that tend to hold heat in the supercold environment of the upper atmosphere. To see how the sun's activity level, as measured by the number of sunspots present, helps determine how many cosmic rays reach Earth, see Figure 27. Also note how closely the cosmic ray density and percentage of low cloud cover correlate.

*Fig. 27* **A. COSMIC RAYS AND SUNSPOT NUMBER COMPARISON**

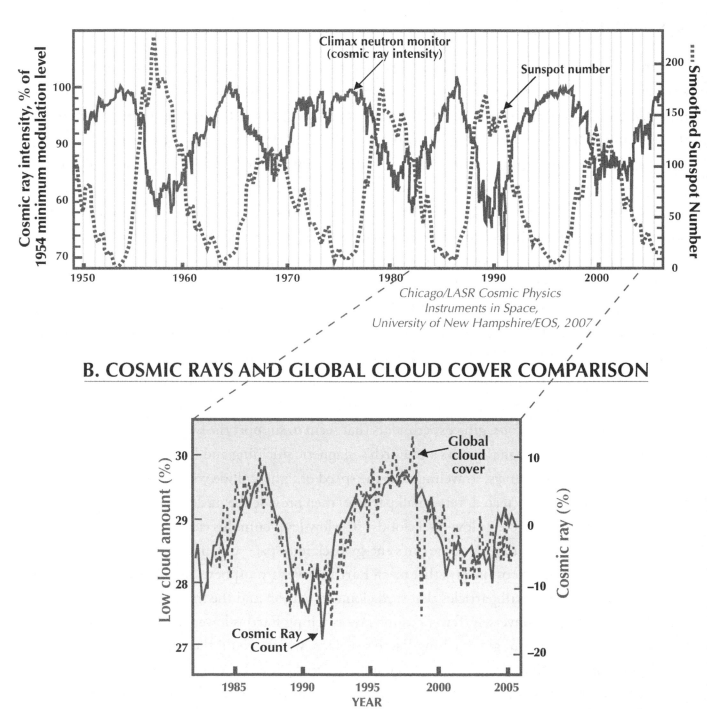

*Chicago/LASR Cosmic Physics Instruments in Space, University of New Hampshire/EOS, 2007*

**B. COSMIC RAYS AND GLOBAL CLOUD COVER COMPARISON**

*H. Svensmark, 2007*

Fortunately, cosmic rays that do enter Earth's atmosphere leave both modern and ancient footprints or indicators to tell us of their relative abundance. Here the excitement builds, since the old footprints tell us, so far, that the volume of cosmic rays reaching Earth was unusually high during times when Earth was experiencing major glaciations. When more cosmic rays are present, the production rate of atoms of the stable isotopes beryllium-10 and carbon-14 rises and allows researchers to measure the number of these atoms that are in Earth's library. An anomaly in this otherwise consistent picture has been that the Cretaceous period that was originally thought to be ice-free was generally a time of active cosmic ray bombardment. Recently, however, evidence of glaciation in Australia during the early Cretaceous Period has been discovered, so the gross correlation is still possible although weaker than during other ice ages.

One reason for the excitement regarding the cosmic ray-cloud cover connection is that no one driver identified to date has stood the test of being able to explain all the major changes in Earth's paleoclimate. Although no one expects to find a single do-it-all driver, everyone would like to see another driver that might explain some of the major climate changes. Will cosmic rays withstand more detailed scientific scrutiny? Will evolution of some of the life on Earth be tied to cosmic ray influx as Dr. Svensmark suggests? Only time and more research will tell.

A major research facility to test the hypothesis that cosmic rays can help cause the formation of clouds is being constructed at Europe's world-class particle physics laboratory called CERN in Geneva, Switzerland. Unfortunately, it will be two or three more years until significant results can be expected from this still-in-construction facility. In the meantime, more research will be undertaken to try to document the times on Earth when cosmic ray volumes were abundant or sparse and compare those abundances with the temperatures at those times.

Let's assume that cosmic rays are shown to have a measurable effect on global climate. Can we hope to predict either when more cosmic rays will appear from space or when the solar shielding will be strong or weak enough to affect how many rays survive to Earth's lower atmosphere where they might create more clouds? The answer to the abundance from space is "probably not" for a specific time because how can we predict when a cosmic-ray generating supernova will occur "close by"? How can we see the rays coming if they are traveling near the speed of light? We are able to predict when our solar system which is part of the Milky Way galaxy, will orbit through one of the spiral arms of the Milky Way, where these massive suns (stars) are more abundant and where more supernovas would be expected to occur. We are still in the vicinity of one of these more star-dense spiral arms, the Orion Arm, and we are in an extended period of glaciations except for our temporary, warm, interglacial paradise. Relative to predicting the sun's future magnetic-shielding activity, we are not yet able to forecast large swings in magnetic activity, but more observations and research may give us the ability to make some meaningful predictions.

What if we find we cannot predict the periods of cosmic ray intensity variations or of the sun's major magnetic cycles? We should then spend our research time and dollars in determining what we should do to best react to any such changes that might occur. In my opinion, we should be spreading our research efforts and dollars studying all of the causes and effects of all the climate drivers, not just focusing on the as-yet-to-be-proven climate-changing effect of carbon dioxide.

Is it likely that one "silver bullet" will be found that controls all observed climate changes? I don't think so, because of all the drivers make some contribution, and any one, or any group of drivers, can have a modifying effect on the climate at any particular time and in either a positive or negative direction. We probably have progressed only part way down one aisle of the very large Earth library of paleoclimate history.

## Driver Commentary

What do you now think is causing the climate to change? Is it still just carbon dioxide? No. Is carbon dioxide important as a temporary retainer of radiated heat? Yes, particularly at low concentrations (Figure 16, page 23). Are there other factors we should take into account and study besides carbon dioxide that we might be able to influence in some favorable way and at some reasonable, nondestructive pace? Probably. I will mention some possibilities in later chapters.

# Chapter Four
# CLIMATE INDICATORS

This chapter addresses your reasonable question: "How in the heck do you scientists know what happened, say, 430 million years ago?" If you really want to understand how we know these things, read on. If you are very trusting of the great numbers of scientists who have tested these methods for their accuracy and interactions and want to get back to more discussion on climate change, just browse on.

Indicators paleoclimatologists use to determine ancient climates and when they occurred can be found in Figure 28. A paper clip for easy referral to that summary page might be useful.

To be able to use in a meaningful way what a particular climate indicator or suite of indicators is telling us about an earlier climate, we must know the time at which the particular climate was occurring. The most common methods used to determine dates rely on the decay rates of certain naturally occurring radioactive elements that were formed at the same time as the medium in which the minerals are found. Each type of radioactive element has its own specific half-life, the time it takes for one half of the original number of radioactive atoms of an element in a sample to decay. Knowing this half-life and measuring the number of radioactive atoms of an element remaining in a sample allow a determination of how long it has been since the element in the sample first formed. With this knowledge, we can date the particular rock or sedimentary layer from which we derive our atmospheric and climate information. Radioactive elements commonly used to determine the dates of geologic materials are shown in Figure 29.

## Stable Isotopes

*This is a biggie.* The best current method we have to pry useful information from the many indicators is stable isotopic analysis. It is also providing answers to what we most want to know, which is what the temperatures were at the time and how much $CO_2$ and oxygen were present, plus the relative amount of measurable trace gases or cosmic rays that impacted Earth's atmosphere. Having said that I have attempted to write a book about global climate change and the factors that affect it for the layman on the street, it is time for a short summary on a highly technical area regarding stable isotopes.

*Fig. 28*

# CLIMATE INDICATORS

| INDICATOR | WHAT WE LEARN FROM IT | ITS LIMITATIONS |
|---|---|---|
| Thermometers | Exact temperature at that site. | Available less than 200 years worldwide. |
| Ice cores | Temperature, $CO_2$, and methane. Dust and sea-salt volumes for wind velocity. | Covers a maximum of 850,000 years at the south pole. |
| Tree rings | Hot vs. cold and moist vs. dry periods and specific age dates. | Only applicable up to a few thousand years. |
| Lake and bog cores | Plant and pollen, sediment variation, and dust (wind velocity). | Rarely find good samples beyond 25,000 years. |
| Ocean cores | Plant, pollen, shell type, dust abundance, latitudes of glacial sediment, volcanic activity, and extraterrestrial impacts. | Ocean floor erased by continental drift (subduction) beyond 175 mya. |
| Glacial footprints | Distribution and timing of glacial or sheet-ice on Earth. | Availability of glacial deposits or striation marks. |
| Vegetation and pollen | Distribution of tropical to arctic trees, shrubs, and grasses. | Only availability of samples to examine. |
| Paleoshoreline positions | Inundation of continents, seafloor spreading rates and sea level. | Availability of widespread samples to examine. |
| Coral type and distribution | Temperature implications of where corals grew and paleoshoreline locations. | Only availability of widespread samples to examine. |
| Paleosoils (Paleosols) | Vegetation type, moist or dry soil type, atmosphere and water chemistry. | Availability of widespread samples to examine. |
| Fossils | Implications of warm or cold fossil types and age environment. | Only availability of widespread samples to examine. |
| Rock, sediments & stalagmites | Environment of deposition, e.g., desert evaporites or tropical rain forests. | Only availability of widespread samples to examine. |
| Stable isotopes | Paleotemperature and atmospheric and fluid composition from many of the above. | Accuracy sometimes affected by changes postdating mineral formation. |
| Archeology | Climate implications of where, when, and how our ancestors lived. | Only robust sites to examine. |
| Written records | Eye-witness climate observations. | Writing only available for last 5,000 years. |
| Atmospheric sampling | Exact amount of gases and particulate matter in the atmosphere. | Only pertains to the last couple of centuries. |

*H. Leighton Steward, 2007*

*Fig. 29*

# NATURALLY OCCURRING ELEMENTS
# COMMONLY USED IN AGE DATING

| PARENT ISOTOPE | DAUGHTER ISOTOPE | USEFUL BACK TO | USED FOR DATING |
|---|---|---|---|
| Carbon–14 | Nitrogen–14 | < 50,000 years | Carbon bearing materials |
| Thorium–230 | Radon–226 | <400,000 years | Corals |
| Potassium–40 | Argon–40 | >100,000 years | Basalts |
| Rubidium–87 | Strontium–87 | 100 myrs | Granites |
| Uranium–235 | Lead–207 | >100 myrs | Various Rocks |
| Uranium–238 | Lead–206 | >100 myrs | Various Rocks |

**The known radioactive decay rates of the half lives of the isotopes shown above allow calculations to be made using the remaining radioactivity to determine the age of the material being examined. Without this ability, our estimates of the age of old rocks, etc., would be based solely on fossil evidence (if any) and result in only a gross approximation of the relative ages.**

All of the atoms of a particular element have the same number of protons in their nucleus. Atoms of elements can, however, have a different number of neutrons that then yields a different mass or mass number. The same basic element but with a different mass number is called an isotope of that element. For example, the most common form or isotope of oxygen is $^{16}O$, which has a mass number of 16. Another less common form or isotope of oxygen is $^{18}O$, which has a mass number of 18. These isotopes come from the same element but are found in different proportions to one another depending on what environmental conditions exist at the time of their precipitation into the solid or fluid being formed.

$^{18}$O is heavier than $^{16}$O. Both are present in water vapor, rain droplets, ice, or snow. The isotopes form the same chemical compounds but because they have slightly different physical properties, such as their melting and boiling points, the heavier isotope, $^{18}$O, is favored in the condensation phase and, as you might expect, $^{18}$O is precipitated out of the atmosphere to a greater degree than the lighter $^{16}$O. A change in temperature or an effect from the distance traveled after the isotopes were formed can cause the ratio of $^{18}$O to $^{16}$O to change. As the temperatures continue to drop, more of the $^{18}$O drops out and further reduces the ratio of $^{18}$O to $^{16}$O. Temperature has the dominant effect, but there is the effect from the distance traveled in the atmosphere. Thus, using other knowledge of temperature and conditions at that time, you sometimes can infer approximately how far the moisture may have traveled before it was precipitated in the form of rain, ice, or snow.

Still other elements have isotopes that can be used by studying various ratios of their isotopes and comparing the effects of current, known factors to the ratios that exist in ancient samples of fluids or solids. This is a marvelous tool that has been used for several decades but is just in the dawn of its application for paleotemperatures and atmospheric makeup relative to the vast amount of material in Earth's library. Isotopic analysis provides information on a large array of minerals and the temperatures that existed at the time of their formation.

## Ice Cores

Detailed ice-core analysis covering the last 400,000 years has provided an excellent record of paleotemperatures for that period. Less reliable but fair climate resolution can be obtained from ice cores that date back to around 650,000 years ago, and gross estimates are possible from studying ice cores dating to 850,000 years ago. As the annual layers of snow and ice get more compacted and eventually folded and squeezed horizontally, the timing of exactly when a layer in the ice was laid down is less accurate. In addition to stable-isotope measurements that provide the paleotemperature values, the cores provide overall atmospheric content for oxygen, carbon dioxide, methane, nitrous oxide, ozone, sulphur dioxide, and other trace gases or minerals. The volumes of dust also can be measured to help infer how windy the atmosphere was and to get a good measure of the general moisture content of the atmosphere. The snow and ice buildups from Antarctica and Greenland have preserved a climate record that can be analyzed in ways that would have been only a wild dream 50 years ago.

The annual snow layers are extremely well preserved for the last 10,000 years and can be individually counted and analyzed. As the snow gets further buried and the snow compacts and becomes ice, the ability to isolate annual layers, even in undisturbed ice layers, diminishes. Good estimates of total layers within an interval can be made, however, and generalized interpretations can be made for periods that are accurate to from a few hundred to ultimately a few thousand years of when layers were laid down in Earth's earlier history.

Being able to calculate the paleotemperatures and also the greenhouse gas, moisture, dust, and trace element content, is a researcher's dream. Unfortunately, there have been no cores taken from undisturbed ice-bearing localities for times earlier than 650,000 years, although an attempt is currently planned in Antarctica that may encounter ice that is a million years old. Let's hope it is undisturbed.

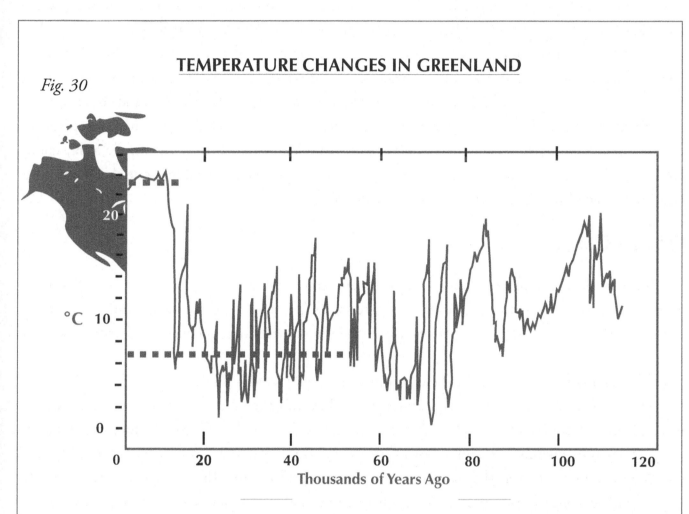

**TEMPERATURE CHANGES IN GREENLAND**

*Fig. 30*

**These temperature changes recorded in ice cores from central Greenland demonstrate the rapid climate shifts that have occurred during the last glacial stage and into the beginning of the current interglacial warm period (Paradise!) The average change (dotted lines)was about 16 degrees Celsius or 30 degrees Fahrenheit. While the change at lower latitudes would have been less, the colder changes and effects on growing seasons would have been disasterous.**

*Modified after L. Gerhard, 2001 AAPG Studies in Geology #47*

Without this record of the astonishingly rhythmic oscillations of the warm and cold periods every 100,000 years and the frequency and magnitude of the many smaller climate changes in between that correlate with Earth's tilt and wobble influences, the uproar over mankind causing all of the current "global warming" would be much louder. Since humans had no influence on these earlier cycles, such as contributing carbon dioxide or methane, it makes it more difficult to argue that human's input of $CO_2$ has become the <u>primary</u> driver of any climate change, a conclusion promoted or at least implied in most media stories today.

As mentioned earlier, the greatest shock that has come from detailed examination of these ice cores has been that $CO_2$ and methane changes followed or lag the temperature changes. Another great shock that came from analyzing these cores was the fact that not only had the climate changed relatively often, but the climate had sometimes changed rapidly and the magnitude of the changes were frequently large. Some of the changes occurred in a few years to a few decades and changed mid-to-high latitude temperatures by many degrees Celsius. See Figure 30 for the record of temperature changes in Greenland over the last 115,000 years. A sudden change of this magnitude today could have a devastating effect on civilization as we know it. No one can possibly argue that these very large and rapid climate changes were caused by humans.

## Ocean Cores

Continual layers of information-bearing sediments can be found that go back to about 175 million years ago. Beyond that time, no sea-floor record exists, because all of the older sea floor has been subducted through the processes of plate tectonics, sea-floor spreading, or continental drift (described in Chapter Three).

The information contained in these continuous layers includes fossils of plant and animal life that existed at that time, the amount and type of dust and the land source from which it came, inferred wind velocities, old glacial deposits that identify cold periods in Earth's history, earlier ocean temperatures and chemistries, and sediment type and grain sizes that can infer whether the sediment was deposited by rivers near the coastlines or deposited in the ocean. The ocean cores are extremely important sources of an uninterrupted sequence of information on paleotemperatures, paleomarine environments, and paleoatmospheric content.

## Lake & Bog Cores

The sediment layers found in old lakes and bogs contain plant and plant pollen remains, shells, animal remains, dust, and various sediment types. The analyses of these contents reveal information similar to that found in ocean cores except that the source of the sediment layers is more localized to the lake or bogs whereas some deep-sea sediment contents have been transported over long distances by submarine currents or drifting glaciers. The oldest cores from current lakes or bogs normally date back only 25,000 years, although one bog deposit has yielded a core with sediments dating to 40,000 years ago.

# Glacial Footprints

As continental glaciers began to melt at their lowest-latitude position, they deposited lot of rock debris that had been ground up and incorporated into their base layers as they slowly moved down-slope. These southernmost positions (northern hemisphere) called terminal moraines thus tell us of the extent of worldwide glacial coverage during that particular glacial period.

Where glaciers move to the coastline and "calve" or break off into the ocean, as they do whether the glaciers are either in a growing or melting phase, they dump a lot of debris where the then icebergs begin their melting in the ocean. When layers of such glacial debris, including boulder-sized pieces, are found on the sea floor at very low latitudes, a reasonable inference can be made that it was a very cold time in the ocean and atmosphere to have allowed so many of these icebergs to float to such low latitudes without having already melted. During very cold periods, glaciers could have formed on mid-latitude landmasses themselves and those glaciers also leave footprints.

As the glaciers move slowly across the land surface, carrying all of this very abrasive rock debris, they leave recognizable scars or striations on the underlying solid rock formations. These footprints of glaciation are often the only visible record that remains of colder periods that occurred hundreds of millions of years ago, because, as mentioned earlier, the sea floor record goes back only 175 million years as a result of the subduction process, and much of the very old rock formations are buried under younger deposits. The oldest-known glacially created striations that have been found on the surface of rock formations are more than 2 billion years old.

# Paleosols (Paleosoils)

Old soils are usually indicative of the climate in which they were formed. The soils contain the remains of plants, trees, insects, and minerals, and the general chemistry of the immediately underlying or adjacent rock or sediment. Simplistically put, soils created in cold climates contain remnants of cold plant types; dry climates contain assemblages of dry plant types, and so on.

Isotopic analyses of the contents of these old soils often yield many clues needed for reliable paleoclimate reconstructions. An article in *Science* by Montanez, et al., 2007, contains detailed estimates of atmospheric $CO_2$, sea level, and also sea-surface temperature over a time from the late Carboniferous to the Early Permian. The more closely spaced data points reveal more detail than the smoothed curves presented in my generalized Grand Summary which were based on 10-million-year sampling intervals and provide some of the evidence that is missing in the generalized curves (see Figure 4, page 2, and Figure 31, page 58). As additional research is performed on the intermediate intervals, more details of the climate changes will be revealed.

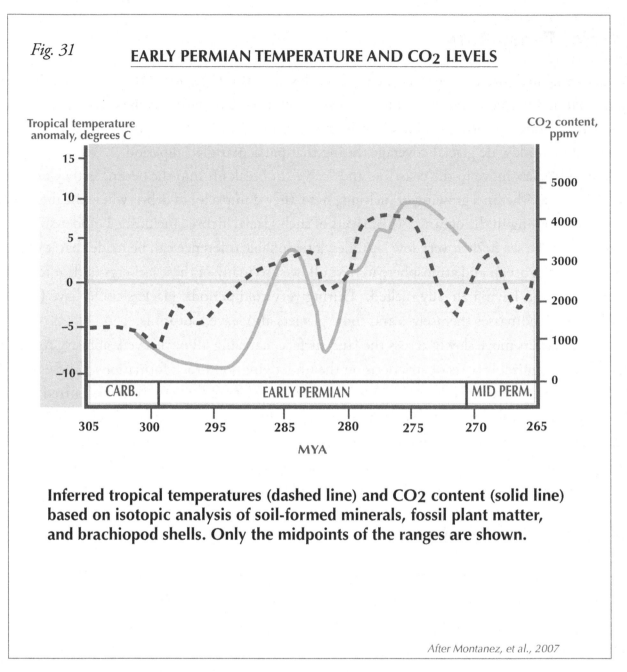

**Fig. 31**    **EARLY PERMIAN TEMPERATURE AND CO2 LEVELS**

**Inferred tropical temperatures (dashed line) and CO2 content (solid line) based on isotopic analysis of soil-formed minerals, fossil plant matter, and brachiopod shells. Only the midpoints of the ranges are shown.**

*After Montanez, et al., 2007*

## Fossil Assemblages

The observation that fossils of earlier life only persisted in a particular morphology or structure for finite periods of time was the first way rock assemblages were successfully correlated over significant distances of Earth's surface. The formations were judged to be time-equivalent if they contained identical fossils or fossil assemblages. Later, high-tech age-dating methods proved that this was indeed a proper application of the fossil information.

Since the animal and plant fossils changed over time in their morphology or structure, using fossils along with the superposition rule that younger sedimentary layers always overlie the older rocks beneath, allowed geologists to reconstruct the rock sequences. Unless the rock layers had been contorted by collisions between drifting continents or oceanic plates or other tectonic forces, these methods provided for the construction of reasonably accurate geologic maps.

The exceptions to the superposition rule were where younger volcanic, high-pressured, molten rock were injected into or between older rock formations and where thrust faults that cause one side of a fault to move both horizontally and vertically had pushed older rock along a fault plane and then up and on top of younger rock. This gave the early mappers a lot of trouble in understanding what they were seeing. Some very large layers of rocks have even been folded completely over where the collision of continental or oceanic plates have compressed these layers, into much shorter, contorted shapes.

As observations of fossils became more extensive and refined, it was found that many fossil types only occurred in a particular environment. For instance, planktonic foraminifera lived within the upper level of the oceans, and benthonic foraminifera lived at greater depths and at various levels near or in the bottom sediments of the oceans. Accordingly, if you found only planktonic foraminifera in the sediment layer, you could infer that the depositional environment was fairly shallow, since no deeper-dwelling benthonic forms were present. On land, certain beetles moved back and forth as the climate warmed and cooled, and some of the body parts of these temperature-sensitive insects were commonly preserved in the rocks, just waiting to tell us a story about the old climate.

## Rock Type

Almost all sedimentary rock gives an indication of the environment of deposition at the time they were deposited. While some river and stream deposits have similar depositional geometrics regardless of the climate in which they were deposited, other rock types, such as extensive reef formations, generally indicate deposition in shallow coastal waters or shallow inland seas that were located at mid-to-low latitude positions where the water temperatures were relatively warm. Evaporite formations containing salt, anhydrite, or gypsum deposits indicate an arid (hot and dry) environment of deposition.

Certain fossil assemblages that are present in the various rock formations give researchers additional hints as to what the climate may have been at the time the rocks were deposited. Also helpful are cores taken from wells drilled to find oil, gas, coal, or mineral deposits. Such cores can be observed and analyzed when brought to a laboratory and they can tell us the history of formations now buried thousands of feet below the surface of Earth.

Since, as you saw in Chapter Three, Earth's albedo or heat-absorbing potential has been identified as important in determining the climate, a sandstone found buried between some evaporate deposits conjures up images of a desertlike area of relatively high albedo that would have reflected a lot of sunlight, while a sandstone that contains tropical plant remains would be interpreted to be deposited from a river coursing through a low-latitude forest. This would be indicative of an area of very low albedo that would have absorbed most of the heat coming from the sun.

Remember that stable isotope studies of some of the minerals found within these sedimentary layers also provide information that tells us of the characteristics of earth's old atmosphere and climate.

## Stalagmites and Stalactites

A somewhat unusual rock type is precipitated in caves when rain water or melted snow enters the ground and trickles down through cracks or permeable soils and drips from the ceilings of caves to form solid "rocks" called stalactites on the ceilings and stalagmites on the floors of the caves. As the water trickling over the surface of the stalagmites and stalactites deposits some of the dissolved minerals the water has picked up on its journey to the cave, in addition to the elements already in the raindrops, the isotropic ratios of some of the various elements provide a record of the climate in which the water or melted snow had been when it entered the cave system. For instance, a high ratio of $^{18}O$ to $^{16}O$ would indicate a warm climate a low ratio of $^{18}O$ to $^{16}O$ a cold climate. The amount or rate of precipitations of the layers on the cave "rocks" also would be indicators of the wetness or aridity of the climate. These growing cave "rocks" can exist for long periods of time and can furnish valuable information from Earth's paleoclimate library.

## Atmospheric Sampling

Today, we can measure the exact temperature and amount of the various greenhouse gases, oxygen, sulphur dioxide, particulate matter, and moisture content of our atmosphere. While this is quite exciting, the time interval of the sampling has been so short that we haven't had time to observe any long-term trends to help us evaluate the causes and effects of precise changes in the atmosphere.

The very recent trends of rising temperature and rising $CO_2$ levels do appear to track each other somewhat, but there were many times in the not-too-distant past when these trends did not tract each other (see Figure 18, page 25 and Figure 22, page 37). Keep in mind that many reputable scientists, including H. M. Fischer, et al., 1999 and U. Siegenthaler, et al., 2005, who have exhaustively studied most of the ice cores, say **earth's temperature changed BEFORE the $CO_2$ levels followed suit**. In fact, after temperature levels began to rise at the end of the Little Ice Age, around 1715, Earth has had some cool stretches that included the "Here comes another ice age" period of the 1940s to the 1970s when glaciers began to advance again in the northern hemisphere and the rising $CO_2$ levels did not track the temperature trend.

I am not advocating that scientists stop studying the present and past impact of $CO_2$ on climate change, just that they also study all other key factors that may be very important in climate change. Try to keep an open mind when someone says climate change is all driven by $CO_2$. Always regard the source of the information. Give most credence to facts from researchers who publish in peer-reviewed journals such as *Science* and *Nature* and not those that shape their message in order to scare the money out of you.  But even in these publications, you will find differing conclusions based on the same basic data.  The science is not settled. ? |

In the meantime, since we don't know all the answers, I recommend that we work toward lowering our greenhouse emissions.  I just don't recommend a single focus on $CO_2$ to the detriment of studying other climate drivers or others more pressing problems that humanity now faces.  More on this later.

## Tree Rings, Pollen and Leaf Shapes

The variation in width of the growth rings of a tree can be used as a fairly good proxy for the general temperature and moisture record for a given year.  Trees grow faster in warm, moist years and slower in cold, dry years or periods.  The more trees you sample, the better the confidence in the outcome.

Unfortunately, trees don't live very long, a few thousand years at best.  But if you can find some nearby buried trees that also live a long time, some of which overlapped the lifetime of your live tree, you can probe a little further into the past with some confidence.

For still older buried trees with no overlap, you can get an idea of their age back to about 50,000 years through carbon 14 age dating and tell by the configuration of the growth rings whether the climate was relatively stable and cold or relatively stable, warm, and moist or some fluctuations in between.  Tree rings provide additional answers to another piece of the paleoclimate puzzle.

Fossil leaves or their imprints show up throughout the sedimentary record.  The shape of leaf edges has been found to indicate whether the tree was growing in a warm or cold climate.  Smooth margin leaves are found in the tropics, and jagged margin leaves are found in the colder climates.  This correlation is quite strong and gives us still another clue to paleoclimates.

The hard case in which the fertilizing part of the pollen is contained often survives burial and is fossilized.  These tiny cases can be identified and traced back to the types of plants from which they came.  Since individual plant types usually grow within specific climate environments, the fossilized pollen are very useful in determining the general paleoclimate at the time the sediment containing the fossils was buried.

## Corals

Like different plant species, some corals live only in very warm water, while other coral types can thrive in cooler waters.  As scientists cross-correlate certain paleocoral types with other paleoclimate indicators, the confidence improves in using various corals as good indicators of regional paleoclimates.

Organisms that build corals secrete calcareous skeletons or structures ($CaCo_3$) so measurements can be made of the isotopic ratios of $^{18}O$ to $^{16}O$, and $^{13}C$ to $^{12}C$ that can be used to estimate the climate at the time the coral was living. Thus, corals provide another indication of relative climate, and if enough samples can be obtained worldwide, researchers can make global paleoclimate estimates.

Most corals grow in a limited, shallow water-depth range. This fact lets us use corals to establish old sea-level positions. Corals normally indicate a location along a shoreline or shallow shelf, although they also can be indicative of a location on an island or atoll. Very high average-sea-level positions generally indicate a time when a lot of water covered the interiors of the continents. These inland seas helped moderate the local climates and reduce the expanse of the deserts. Low average-sea-level positions often indicate a climate that should have been somewhat harsher and drier as you move inland from the coastlines.

Other factors, such as increased or decreased sea-floor spreading rates or tectonic uplift or subsidence can also affect paleoshoreline positions independent of the climate at that time. As spreading rates increase and more hot, molten "rock" rises to or near the surface of the sea-floor, the large and very long ocean ridges bulge upward and displace more seawater onto the continental margins. When spreading rates decline, the ridges cool, shrink, and subside, and sea level falls. Again, all pieces of the puzzle need to be examined before a reliable cause-and-effect analysis can be presented with some confidence.

## Paleoshoreline Positions

As mentioned in the discussion on corals, paleoshoreline positions indicate the relative sea level at any given time, including the times when the shorelines moved off the continental shelves of the oceans and into the interiors of the continents. Knowing these shoreline positions, when combined with other indicator data, can help with a better interpretation of what was occurring at a particular time that may have had a directional influence on the climate. For instance, if sea levels were unusually high on a global scale but there was evidence of glaciations, the swelling of the mid-ocean ridges caused by accelerated sea-floor spreading rates would be a likely culprit. This could suggest a time of more $CO_2$ being emitted into the atmosphere from the spreading centers or from the more active subduction zones and their related volcanic activity.

While a time of no glaciation also can cause high sea-level stands, there is usually a limit on how high the melted ice volumes can cause the sea level to rise. Alternatively, very low sea-level stands could indicate a time of very low mid-ocean spreading rates if little evidence of glaciation exists. Low spreading rates allow the magma extruded at the spreading center to cool and sink and not be replaced so quickly with a bulge of large volumes of more hot magma.

## Thermometers, Written Records and Archeology

These indicators speak for themselves and are sufficiently covered in the Climate Indicators Summary in Figure 29, page 52.

## Chapter Five
# EXPANDING ON THE GEOLOGIC PERIODS

This chapter will offer some modest expansions of the descriptions of the major geologic intervals that are summarized in Figure 4, Page 2. As with the temperature, $CO_2$, and oxygen curves that are generalized because of the exceedingly long time between data-collection intervals, these additional observations are also generalized.

When more data becomes available, I am certain these curves and generalizations will be filled with the same constant climate variability that we now know has occurred in the more recent past and that is continuing today. As you read these expanded summaries, you will find it helpful to refer frequently to the overall summary charts mentioned above.

## PRE-CAMBRIAN EON

### Archean Era, 4.6 billion to 2.5 billion years ago

The sun and Earth and other planets formed about 4.6 billion years ago. About 50 million years later, most astronomers believe a Mars-sized body collided with Earth, and the ensuing debris formed the moon. Earth's surface was initially molten, the sun's energy was 30 percent less than today, and the moon was much, much closer to Earth (less than 20,000 miles away), resulting in tremendous tidal effects on Earth's molten surface. As Earth gradually cooled, some solid rock began to form and clump into the earliest "continents" which floated on the molten surface. Space within the solar system was filled with debris, and these "rocks" of all sizes bombarded Earth very frequently which helped keep Earth's surface hot.

After about three-quarters of a billion years, the frequency of the extraterrestrial collisions dropped sharply, and Earth continued to cool at a more rapid pace. The early atmosphere, which contained no oxygen, was filled with carbon dioxide, nitrogen, and some water vapor and, by about 3.8 billion years ago, probably saw water precipitation and the formation of early oceans. The tremendous tidal effects from the still-close moon, resulting in tides of several hundred feet would have been unimaginable. Tides and waves would have been constantly eroding any earlier continents that had formed. Analyses of both oxygen and silicon isotopes have provided data that some researchers say indicate that ocean surface temperatures were quite high for much of the time between 3.5 and 2.5 billion years ago, but other researchers claim to have found evidence for glaciation (extreme ice cover) before the end of this era. The time intervals are so long that both may be correct.

A pivotal thing happened about 3.5 billion years ago with the occurrence of cynobacteria, the earliest form of life currently known from studying Earth's history. As this life flourished over the next one billion years, it used the sun's energy to sustain its own life and released oxygen into the atmosphere. That ultimately paved the way for the great oxidation event of 2.4 billion years ago. The very hot environment that gave rise to the cynobacteria still can be found in places on Earth's surface, such as in the thermal pools in Yellowstone Park, where cynobacteria still thrive in waters of 160 and 200 degrees Fahrenheit!

Best estimates, based on data from a few of the remaining ancient rock outcrops, suggest that several continents of modest size were present south of Earth's equator with three or four much smaller ones scattered elsewhere (see Figure 4, page 2).

## Proterozoic Era, 2.5 billion to 540 million years ago

Many things of interest occurred during this next long era of Earth's history. The atmosphere finally contained breathable oxygen, photosynthesis began, and the first multicelled algae, early aquatic plants and some fungus appeared, and probably two periods of glaciation.

Best estimates of measurements of $CO_2$ and oxygen using stable isotope analyses indicate that $CO_2$ levels near the end of the era were almost twenty times current levels and oxygen content was slightly lower than today. Sea-surface temperatures, as indicated by stable isotope measurements, cooled to 20 to 30 degrees Celsius by 1.5 billion years ago. Current average near-surface global temperature is 15 degrees Celsius or 59 degrees Fahrenheit.

Cooling of Earth took a dramatic step around 850 million to 650 million years ago in that indications of significant glaciations, possibly three or more separate ones, shows up in the geologic record. The length of the glaciations that left the characteristic scratches or striations on the rocks is not well known, however, and the $CO_2$ content and sea-surface temperatures during or immediately prior to the glaciations is not well known, either. I assume lower sea-surface temperatures existed at the time of the glaciations!

Cooling of Earth also had helped create more solid landmass and a large continent had coalesced near the equator by around 2.0 billion to 1.5 billion years ago. By the time of the just-mentioned glaciations, however, the large continent had broken into many smaller continents, all of which show some evidence of continental glaciation. Finding evidence of glaciation on all continents supports the idea that a large portion of the surface of Earth may have been encountering glaciation, which has given rise to what many refer to as the "Snowball Earth" period.

# PALEOZOIC ERA

## Cambrian Period, 540 million to 490 million years ago

Earth warmed again by the beginning of the Cambrian Period. The $CO_2$ levels were very high (18 times higher than today), and oxygen levels were also high, near to those oxygen values on Earth today (see Figure 4, page 2). Temperatures probably stabilized somewhat due, in part, to the greatly diminished frequency of extraterrestrial impacts.

The single most impressive event of all the eras lies in the explosion of life forms and the abundance of life that occurred during the Cambrian. While some forms had begun to appear near the end of the Proterozoic Era such as worms, jellyfish, corals, and sponges, their abundance grew dramatically during the Cambrian, and new life forms appeared, including clams, snails, squid, octopi, trilobites, insects, spiders, crabs, shrimp, and early fish. All of this happened in only a 50-million-year period whereas the earlier eras each covered a couple of billion years.

A very large continent, Gondwana, had formed in the Southern Hemisphere and was surrounded to the east, west, and north by much smaller continents.

Near the end of the Cambrian, oxygen and $CO_2$ levels dropped, and many of the new life forms in the sea became extinct, while others began to acquire a different morphology (internal or external body forms) to survive the atmospheric and upper-ocean chemistry change.

## Ordovician Period, 490 million to 440 million years ago

The average temperature began quite warm, and $CO_2$ levels, while lower than in the Cambrian, were still high relative to succeeding periods. Oxygen levels dropped to very low values during the early and mid-Ordovician and then began a steady increase that would continue for more than 60 million years. The increase in oxygen possibly helped oxygenate the near surface oceans and stimulate a rebound of life that had an abundance of coral reefs, true fish, sponges, and other shelled organisms such as brachiopods and mollusks. Many other new species appeared.

The temperature declined until the Late Ordovician but still ended the period probably several degrees warmer than Earth's temperature today. Near the end of the period, the temperature began a protracted increase that would continue for more than 60 million years

The Appalachian Mountains were being uplifted during the early and middle Ordovician, mountains that have since worn down dramatically from their earlier, near-Himalayan heights.

Gondwana began to drift toward and partially over the south pole, with several smaller continents located in near-equatorial positions.

The Ordovician ended with a large extinction event, possibly from a major extraterrestrial impact.

## Silurian Period, 440 million to 400 million years ago

While $CO_2$ levels dropped to the lowest levels up to that time (but still many times higher than today's levels), the oxygen content of the atmosphere rose to its then-highest level, which stable isotope calculations suggest were several percentage points higher than today's level. The moderately warm early Silurian temperature dropped for some relatively brief intervals into icehouse conditions around 440 to 430 million years ago, then warmed again. The very large continent, Gondwana, was still south of the equator and extended to the south pole. Evidence of the glaciation exists in the rocks that were situated on Gondwana, and some evidence indicates a glaciation extended to as low as 60 degrees south latitude despite general $CO_2$ levels being ten times higher than today.

Life in the sea was very abundant and included huge eurypterids (scorpionlike creatures ten feet long that ruled the seas).

The major life-changing events of the Silurian, however, took place not in the sea but on land. Life moved from the sea to the land, and this included the first land plants. The surface of Earth began to look green for the first time! Just imagine, for more than four billion years Earth's land surface had no plants or animals! The first record of animal life on land was that of scorpion, insect, and spider fossils.

The very high levels of oxygen must have aided the growth of much of the animal life, because many of the scorpions and other insects were huge, much larger than when they lived in times of lower oxygen.

High concentrations of oxygen usually mean a high incidence of fire. The earliest land plants probably paid a price for finally making landfall during this time of elevated oxygen. Evidence of fire is indicated by the presence of large amounts of charcoal in the rocks of this time.

## Devonian Period, 400 million to 360 million years ago

The Devonian began with moderate and rising temperature and $CO_2$ levels, which, by the Middle Devonian, began a long, significant decline that continued into the next geologic period. Oxygen levels, which were very high, declined dramatically during the Early Devonian, bottoming out in the Mid-Devonian at the lowest level since the Pre-Cambrian Eon. This dramatic decline, although rising somewhat in the Late Devonian, may have combined with extraterrestrial impact activity at the end of the period, to cause the large extinction event that has been identified (see Figure 4, page 2).

The first amphibians appeared during the Devonian which meant that, for the first time, non-insect animals that had evolved from the fish had finally made landfall. Also, some flying insects made their appearance. This transition is recorded in the fossil record of that time. The first true sharks and rays had appeared in the seas and they survived this large extinction event and all that have occurred since. No one knows for sure why they were more resilient than most of the other life forms present at that and subsequent times.

Gondwana still occupied the Southern Hemisphere and several of the smaller continental landmasses had begun to clump together slowly near the equator.

Land plants were continuing to expand, and many mosses, rushes, and ferns appeared.

## Carboniferous Period, 360 million to 300 million years ago

The Carboniferous began warm and wet, but the temperature decline set in motion in the Devonian Period continued and helped change the fauna and flora before the end of the Carboniferous. The first trees appeared, and the vegetation was lush in the widespread swamps around the world, probably helping to cause the plunge in $CO_2$. All of this vegetation ultimately would be buried in these vast swamps and form 90 percent of the coal deposits present in the world today. Despite the extensive swamps that had occupied large portions of the now present two huge continents, Gondwana and Laurasia, they both also contained very large deserts in their interiors, a common feature where land is located so far from a coastal area.

Toward the end of the Early Carboniferous, oxygen levels began a dramatic rise. Flying insects appeared that had wing spans of up to 3 feet for the paleodragonflies. Then came the first reptiles. The fossil record indicates that huge "forest fires" occurred at this time, probably triggered by the high and rising oxygen levels.

The two huge continents, Gondwana and Laurasia portions of which were located over the equator, probably helped divert equatorial currents to higher latitudes and caused or aided the onset of glaciation by providing more moisture to the coldest areas of southernmost Gondwana.

Global temperature continued to drop throughout the Carboniferous, and by the Middle Carboniferous, about 325 million years ago, the $CO_2$ content of the atmosphere plummeted below any currently identified earlier levels and almost as low as today's level. This icehouse cycle, with continuing glaciations, persisted for the remainder of the period and throughout most of the next period, the Permian.

## Permian Period, 300 million to 250 million years ago

The very low $CO_2$ levels of the late Carboniferous persisted into the Permian. The icehouse (glacial) conditions also continued throughout most of the Permian. Oxygen levels soared to their highest levels, approximately 30 percent, which is the highest level measured in the history of Earth. That level would be about 50 percent higher than today's level. Even you and I might have been able to finish a 26-mile marathon. Near the end of the Permian, oxygen levels plummeted. Temperatures recovered to modestly high levels, much warmer than today, and the icehouse period began to end.

The huge continents present during the previous period had collided and formed one supercontinent, called Pangaea. For the first time, all life forms that had evolved on land could "communicate" with one another, absent an unclimbable mountain range or impenetrable desert barrier. The deserts were very vast, and that evidence shows up in the geologic record in the form of thick evaporite deposits such as salt, gypsum and anhydrite.

Life on Earth diversified further, with lizards, snakes, and some animal-like creatures, the synapsida, appearing. The best known of the synapsida were the Dimetrodon and its other four-legged, air breathing cousins. Forerunners of the dinosaurs appeared near the end of the period. All in all, this period of very high oxygen levels did not see as much diversification as other periods that were more oxygen-starved. This might suggest that times of oxygen starvation created pressure for the animal kingdom to diversify in order to survive (Ward, P.D., 2006). Meanwhile, the long-term, low-levels of $CO_2$ were not conducive to robust plant life on this supercontinent.

At the end of the Permian, all hell broke loose. Catastrophe, Armageddon, whatever you want to call it, approximately 90 percent of all life on land and in the sea died in a geologic instant of time. Oxygen levels had crashed, and $CO_2$ levels had risen. Other things happened, causative things responsible for this greatest of all extinctions. Most scientists believe a giant asteroid struck Earth in the vicinity of present-day Antarctica, which was then the continent of Australia that was part of the supercontinent called Pangaea. As mentioned earlier, researchers from Ohio State University have recently located a gravity anomaly that has the earmarks of a supergiant impact crater about 300 miles wide beneath the eastern ice shelf of Antarctica. The crater that wiped out the dinosaurs 65 million years ago was "only" 110 miles wide. If the crater is from a giant impact of about 250 million years ago, as is claimed in the initial estimates, the researchers have found the cause of the greatest loss of life in Earth's history.

Concurrently, a vast volcanic field erupted in Siberia that ultimately covered an area the size of Alaska. Either of the events would have killed off a huge amount of life, but having both happen simultaneously would have just made things worse.

Some scientists say there was no extraterrestrial impact, that the volcanic eruptions did it all. I don't disagree that super-volcanic eruptions over a long period of time could have been the killers by causing acid rain, but why the coincidence of other mega volcanic eruptions with these giant impacts? The timing of the arrivals of large asteroids is certainly not caused by volcanic eruptions!

My guess is that both things happened. The shock waves from the impact by an asteroid estimated to have been 30 miles wide, which was many multiples larger in total mass than the one that supposedly killed off the dinosaurs, rolled around the surface of Earth and/or through Earth's interior and caused such violent earthquakes and fissures in the outer crust that the Siberian Traps (volcanic field) on the opposite side of Earth sprang to life or, if already present in a smaller area, probably expanded immediately with the newly formed conduits. Any faults that had pent-up tensions would have had an opportunity to release those tensions as a result of this impact-caused ground roll.

The atmosphere would have been full of toxic acid rain and toxic to life. The oceans would have been poisoned except for the deeper environs. How anything on the surface survived is a mystery in itself. But some did survive! Let's go to the next period and find out not only what survived but also what ultimately flourished.

## Triassic Period, 250 to 200 million years ago.

Oxygen levels were relatively low through out the Triassic. $CO_2$ levels began to fall again following the extinction event but apparently "stabilized" at a modestly low level. Temperatures were high early but dropped significantly near the end of the Triassic. Temperatures were, however, still warmer than those today.

The near-lifeless Earth persisted for a few million years, and then, around 240 million years ago, life began to recover and transform into species that could breathe adequately in the low-oxygen environment. The new life forms flourished, helped by the fact that little competition existed.

In the seas, many new species developed, while many others were modified. New types of coral, mollusks, ammonites, and sea creatures (ichthyosaurs) appeared in the recovering seas. To see what some of these early creatures probably looked like watch some of the programs that cover these ancient periods on the *Science, History,* or *Discovery* channels.

On land, the few forerunners of the dinosaurs developed into multiple combinations of body sizes and shapes. Some of these remnants of the Permian extinction took to the air and became the first flying reptiles (pterosaurs). Many large mammallike creatures existed. But none was a match for the two-legged dinosaurs that had developed a breathing system that allowed them to utilize the small amount of oxygen much more efficiently (Ward, 2006) and the 185-million-year reign of the dinosaurs began. Meanwhile, the earliest true mammals evolved and, because of their diminutive size, escaped extinction at the hands of the carnivorous dinosaurs.

Pangaea, the supercontinent, was initially intact but was beginning to show signs of breaking up as the period ended. A few asteroids of some size struck Earth in the Late Triassic but no large climate change has yet been identified with these asteroids impacts.

Another mass extinction event occurred at the end of the Triassic, possibly caused by a further reduction in the oxygen content of the atmosphere.

## Jurassic Period, 200 million to 145 million years ago

In the early Jurassic, oxygen levels were very low, at the lowest levels recorded in at least the last 500 million years. The temperature had increased slightly by the mid- Jurassic, and so had the $CO_2$ levels.

For the larger animals present, the low-oxygen environment continued to favor the dinosaurs. While mammal species expanded in numbers, they remained quite small in size. Some of the dinosaurs that had developed feathers in the Triassic continued evolving into birds. Other dinosaurs became true giants by the late Jurassic.

Life in the sea, which had been affected by the Triassic extinction event, recovered well and appeared to have less trouble living in a lower-oxygen world than did the larger non-dinosaur creatures on land. All of the sea life with calcareous (limestone-like) shells flourished in a higher $CO_2$ environment because it was partly from the $CO_2$ that the creatures derived their carbonate shells.

The $CO_2$ level, many times higher than today's, apparently did not acidify the oceans enough to prevent the formation of abundant shells for life in the sea, as some scientists fear may happen with a modest increase in $CO_2$ levels in today's seas. However, it is possible that more detailed sampling of the Jurassic sediments may find that sea-life abundance and $CO_2$ levels did generally vary together.

Pangaea began to break up during the Jurassic, and this included the early drifting apart of South America from Africa and North America from Europe. The major environmental impact from this break-up would have been the reduction in size of the great deserts, since the newly formed waterways brought moisture to the interior of the fracturing Pangaea.

## Cretaceous Period, 145 million to 65 million years ago

Moving nearer to the present, much more data is available. Subduction during continental drift has not erased all of the continuous record of the sea-floor sediments as it has for sediments that were older than 175 million years of age.

The Cretaceous atmosphere began with a high $CO_2$ level and a warm, humid climate. While the $CO_2$ level declined later in the period, it was about 10 times higher in the Early Cretaceous than today's level. Several estimates place the average global temperature at about 20 degrees Fahrenheit higher than at present! The oxygen level started out very low but rose throughout the period, concurrent with the reduction in $CO_2$ levels.

Sea floor spreading rates were much higher than at present, and sea level was at least 600 or 700 feet higher than today. Shallow seas covered the interior of many continents, including the central interior of North America. Marine life flourished, and corals and rudistids grew in abundance. The rudistids, or mollusk (clamlike) animals that attached themselves together to form large reeflike structures, dominated the shallow seaways. Outcrops of Cretaceous rocks are full of either macrofossils that you can easily see such as corals and clams, or microfossils that are present in lime mudstones and chalks such as Cocoliths. The famous White Cliffs of Dover are composed of such chalk.

Although the dinosaurs didn't really need the rising oxygen level, they adapted well to it and continued to kick butt in the chow line.

The continents were well scattered and not too dissimilar from today's distribution except that North and South America were much closer to Europe and Africa. No major continental collisions were occurring, no extensive glaciation existed, and, except for the rising oxygen level, the climate and atmosphere were probably about as stable as it gets.

Then nearly all hell broke loose again. An asteroid about six miles across and traveling at 40,000 to 60,000 miles per hour traversed Earth's atmosphere in about 3 or 4 seconds and struck shallow water or land at what is now the Yucatan Peninsula in Mexico. No doubt about this one; it left a crater that measures 110 miles wide! It is estimated to have created an initial wave 3,000 to 5,000 feet high which was still huge as it washed over a slightly submerged Florida, carrying boulders from the impact site out into the western Atlantic. Some tsunami!

Molten debris was lofted into space and fell all around Earth, setting fires worldwide and causing acid rain. Wind with velocities estimated at 600 miles per hour probably blew away huge dinosaurs unlucky enough to be within a hundred miles of ground zero. Just think what havoc the tsunami that swept up the interior of North America caused. The dust, which stayed in the atmosphere for years, darkened the sky and reflected most of the heat coming from the sun above. This turned the previously warm climate into an instant, albeit brief, icehouse.

About this time, the Deccan Traps volcanic field in India became active and expelled great volumes of magma and related gases. This occurrence exacerbated the toxicity of Earth's atmosphere and likely caused many more extinctions.

The dinosaurs and more than 60 percent of all species on Earth did not survive. Bad for them, lucky for us, because, as you will see in the next period, it gave rise to many more mammals, and eventually our ancestors showed up.

## Tertiary Period, 65 million to 2.6 million years ago

The carbonate, limestone-rich, coral-rich, and dinosaur-ruled world of the Cretaceous changed in that fiery instant to a world of mostly clastic (sand) type sediments that were eroding from the suddenly more barren land surface. Many of the plants kept a toehold, however, and soon the world turned green again. The giant sequoias survived since they were growing in the shadow and protection of the mountains that existed in the west. New pine trees proliferated, as did cactus-type plants.

Most of the fish and many of the shellfish made it through the catastrophe with only a temporary reduction in numbers. The alligators and turtles, close relatives of the dinosaurs, survived essentially intact, because both were able to go underground and hibernate for an extended time and sleep through the brief icehouse. When an alligator goes into hibernation, its heart rate drops from thirty beats per minute to two beats per minute and little oxygen is required.

After a few million years, two groups not only survived but also flourished in a big way. Birds evolved into many different types, and the mammals, heretofore small and rather unimpressive, diversified into many new species and became the dominant force on land. Many animals became huge and evolved into today's rhinos, elephants, hippos, and whales. Others species grew larger, more efficient brains, such as the monkeys, apes, and dolphins. Man's best friend's, the dogs and also the cats (some very large), proliferated, although with characteristics probably much wilder and more aggressive than today's household varieties.

The consensus of anthropologists is that our ancestors, the *Homo sapiens*, slowly evolved from the Tertiary Period apes over a period of several millions years. By God, intelligent design, just plain evolution or some combination, our brain sizes increased and developed the ability to reason and plan and be aware of self and others and of how the actions we take today will likely affect tomorrow. We have no peers and are clearly the most dominant force Earth has ever known, even though we lack the claws and teeth of the earlier rulers of their time. The long-in-coming paradise that we live in today is only spoiled by our inability to get along with each other. So we have invented things much more formidable than large claws and teeth. What a pity.

The climate changed a lot during the Tertiary period which ended 2.6 million years ago. Earth's temperature continued a long decline accompanied by lower and lower $CO_2$ values (see Figure 22, page 37). A couple of the brief exceptions were the $CO_2$ spikes following the dinosaur-killing asteroid impact of 65 million years ago and the significant temperature rise 55 million years ago, which is, interestingly, near the beginning of the collision of the Indian plate with the Asian plate. The long-term drop in atmospheric $CO_2$ may have been caused or at least aided by the sudden increase in chemical weathering brought about by this collision which caused a great increase in elevation of the India-Asian area.

The sudden warm spike of 55 million years ago is thought by many to have been caused by a large methane (clathrate) release that caused sudden greenhouse warming. But methane is believed to only remain in the atmosphere about 10 years. Maybe the sudden warming started a chain reaction of positive feedback for warming. Maybe the solar-orbital factors were poised for a shift to a warmer period on Earth, and the methane burst just pulled the trigger. Maybe the sudden and rapid movement of India from near the South Pole to near its location today, caused great rates of subduction and disturbed large deposits of methane frozen in the seafloor. I'm sure the answer is to be found somewhere in Earth's library.

Then, about 35 million years ago, Earth began to cool more rapidly, and the south pole that was located atop the continent of Antarctica developed permanent glacial ice. By seven million years ago, Greenland had developed huge volumes of ice that persisted through the end of the Tertiary and remain there today.

With the exception of the ongoing collision of India with Asia, the rest of the continents were drifting rather freely on Earth's molten subsurface. How important was this one collision from a climate standpoint? High, fragile, rocky surfaces like those in the Himalayas and the Tibetan plateau were providing a constant new source of silicate rock fragments and surfaces that could undergo chemical weathering and lower, or continue to lower, the long-term $CO_2$ content of the atmosphere. As the $CO_2$ levels dropped, so did the temperatures … or was it vise-versa (see Figure 22, page 37)?

## Quaternary Period, 2.6 million years ago to present.

By the beginning of the Quaternary Period, 2.6 million years ago, very large ice sheets were present in the higher latitudes of the Northern Hemisphere. The advance and retreat of these ice sheets have culminated in our current state of approximately 90,000 years of cold, windy, and dry glacial periods followed by rather brief 10,000-year interglacial, warmer, and wetter periods. These warm interludes are then followed by another 90,000 years of cold, windy, and dry glacial conditions.

Three or four million years ago, the open passageway for equatorial ocean currents between the Atlantic and Pacific Oceans was closed off by tectonic uplift and volcanic outpourings. This newly formed land is now the Isthmus of Panama. With the free movement of warm, low-latitude Atlantic waters "suddenly" blocked, the climate patterns began to change, and for the first time in hundreds of millions of years, very large continental ice sheets began to form in the Northern Hemisphere. While Earth had experienced glacial conditions in the Southern Hemisphere over the previous 30 plus million years, the glacial cycles became more frequent and rhythmical in the Northern Hemisphere after the Panama closing.

The timing of the advances and retreats changed during the Quaternary Period. The earlier cycles of glacial to brief interglacial climates were around 41,000 years, and then, about 900,000 years ago, the cycles became about 100,000 years in length, and the ice sheets became even larger. The most recent glacial advance, which began to retreat only around 15,000 years ago, caused ice (from snow) to build up to nearly two miles thick just north of what is now New York (see Figure 32). These facts are real and not being debated. They are, of course, fodder for catastrophe-oriented movies and TV programs. While these ice buildups and relatively rapid switch to a melting phase did occur, the melting and rising sea level took thousand of years, not the instant changes depicted in the movies or some television "documentaries".

What did occur very quickly, in a decade or less, was a flip of the climate from very cold to suddenly warm or to a somewhat lesser extent, vice-versa. Fortunately for our ancestors and relative to our future, the flip to the initial cold periods did not immediately take the climate to the depths of the glacial cycle temperatures. While that is true, as verified by careful studies of ice cores taken in Greenland and Antarctica and from mountain glaciers, the temperature did "immediately" drop enough to cause "instant" havoc to the growing cycle in what are now the mid-latitude bread baskets of the world. From feast to worldwide famine in a few short years is a truly horrible scenario.

Let me cite some real examples of how much even small changes in global temperatures can impact Earth's civilizations. Around 1,350 A.D., the warmth of the Medieval Warm Period began to drift into a somewhat cooler and drier cycle known as the Little Ice Age. This cooling had an effect on global average temperature of only two to four degrees Fahrenheit. This change would have been even less near the equator but would have been still greater in the high latitudes.    The Vikings had made it to North American and established villages surrounded by some agriculture and cattle ranching. But when this rather modest global climate cooling occurred, even this hearty group of explorers froze or starved to death.

The Anazazi Indians, with their fabulous cliff dwellings in northern New Mexico, also disappeared "overnight" during this Little Ice Age and the attendant drought that affected their region. They may have simply moved south, but either the drying and dying of their croplands or the harsher weather was probably what caused them to abandon such fine dwellings.

In Central America, a strong case can be made that the very advanced and unusually long-lived civilization of the Mayans fell on hard times and disintegrated when a climate-related, prolonged drought changed their environment around 800 A.D. This probably forced them to abandon their marvelous cities during the next century.

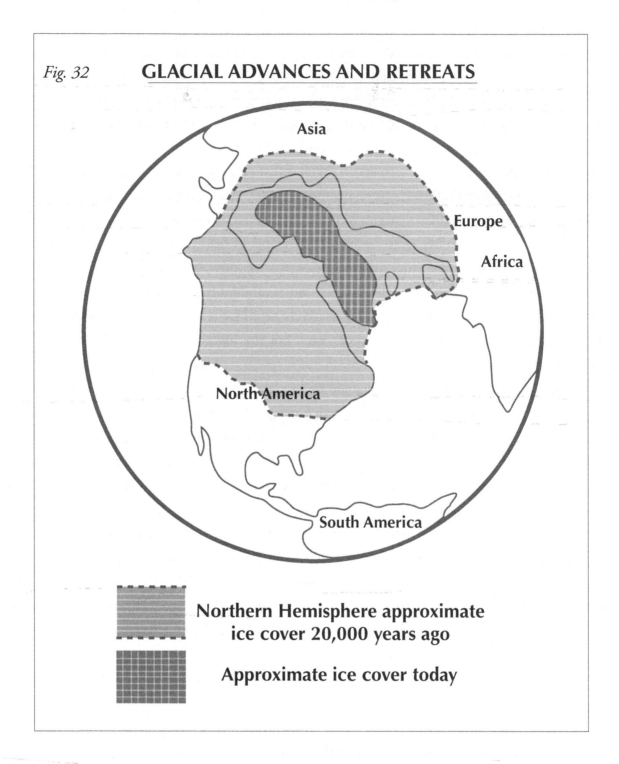

**Fig. 32**  **GLACIAL ADVANCES AND RETREATS**

Asia

Europe

Africa

North America

South America

Northern Hemisphere approximate
ice cover 20,000 years ago

Approximate ice cover today

As the archeologist, anthropologists and paleoclimatologists continue to get together to review earlier civilizations, they are finding many more worldwide examples of the negative or positive effects of climate change on earlier civilizations. In nearly all cases, the dislocations were caused by colder and/or drier climates, not warmer and wetter climates like the one we are living in today.

With the exception of the brief and modest spikes in $CO_2$ during the interglacial cycles, worldwide $CO_2$ levels are as low as they have ever been measured over the span of Earth's history. So what? I know I'm really enjoying this brief paradise of a warm trend in which we are currently living and

I certainly do not want the $CO_2$ levels to drop again below their pre-industrial levels of 280 ppm, especially if they are following the temperature down!

So here we are, basking in one of the very few instances of earthly paradise for mankind. Walk outside and rejoice in its hospitality, whether it is zero or 100 degrees. Over the year, it averages about 59 degrees Fahrenheit worldwide and we can live very comfortably in the temperatures and growing seasons that 59 degrees Fahrenheit provide.

*Chapter Six*
# MORE ON THE GREENHOUSE

## Greenhouse gases are good!

Without greenhouse gases Earth would be frozen solid, its global average temperature approximately minus 18 degrees Celsius or zero degrees Fahrenheit. We are currently in a paradise of a greenhouse mixture and orbital orientation for humanity with this average global temperature of 59 degrees Fahrenheit.

How much variation in greenhouse gases can we tolerate and still be considered to be in a paradise mode? We don't know for sure. We do know that during the most recent glacial extremes, with $CO_2$ levels of only 180 ppm (currently 380 ppm) and methane volumes nearly half what they are today, that Earth was relatively uninhabitable north of about 40 degrees latitude in North America and Europe because of pervasive ice cover. A large number of Earth's people live near or above that latitude today; this includes major cities such as New York, Chicago, Seattle, Paris, London and Moscow. But since we now know that $CO_2$ levels follow, not cause, the temperature changes and that $CO_2$'s ability to trap more heat declines for each additional molecule of $CO_2$, the effect of rising $CO_2$ certainly appears not to be as significant as once thought by nearly all of us and still thought by most of the general population.

If Earth re-enters a cooling phase and the cooler oceans quickly absorb a lot of atmospheric $CO_2$, the greenhouse effect would decrease at a greater rate with each molecule of $CO_2$ decline. To say it another way, the ability of the next declining unit of $CO_2$ to cause more global greenhouse cooling is greater than the next increasing unit of $CO_2$'s ability to cause more global greenhouse warming. This is because of the logarithmic decline properties of atmosphere $CO_2$.

Another detrimental effect of returning to the very low $CO_2$ levels of the glacial cycles would be the drop in global plant growth, not just from the cooler climate and attendant shrinkage of the agricultural growth belts, but in starving the flora of its life-giving $CO_2$. Hence, the additional potential benefit of sequestering $CO_2$ today in case it is needed to try and counter the many effects of global cooling as $CO_2$ levels become very low again.

During the Cretaceous Period, when $CO_2$ saturations were 4 to 10 times higher than today's levels, humans were not around to have their survival skills tested. These much warmer climates did allow certain species of land and sea life to flourish, however. It must have been particularly favorable for the growth of creatures with big teeth and claws! This super-hothouse period occurred 90 to 130 million years ago, when dinosaurs still roamed earth. Measurements of stable isotopes from that period indicate a probable variation from today's worldwide average annual temperature of 59 degrees Fahrenheit, of as much as 20 degrees Fahrenheit or a world wide average day and nighttime temperature of nearly 80 degrees Fahrenheit. Whew! It is little wonder that we find dinosaur fossils in rocks of this age far north of the Arctic Circle (66.7 degrees north latitude).

If we had been present during the Cretaceous and had not been eaten by the big lizards or sea monsters, we might have been able to survive, but except for the ice-free poles, would probably have always been hotter than blue blazes and feeling like we were breathing at 15,000 feet because of to the lower oxygen content of the atmosphere (see Figure 4, page 2).

But, was it just the high $CO_2$ level that created this hothouse period? It could have been, but additional research may find that this part of the Cretaceous was a time of very high solar activity that increased the sun's radiance somewhat or that the sun's magnetic activity spiked and created much more solar wind that shielded cosmic rays from Earth which possibly resulted in much less low cloud cover. Scientists do know the sea floor spreading rates were much higher during the Cretaceous, so more heat from within Earth would have been delivered to the oceans. Or was it some other factor we have not yet recognized? Our descendents will find the answers in Earth's library.

If our climate were to move toward the warmer and wetter than today mode, conditions might not be too difficult to tolerate, since, generally, "warmer and wetter, most life lives better." A move in the opposite direction; "cold and dry, many life forms die."

Would diseases in a much warmer, wetter climate also have flourished? Possibly, unless the greater abundance of plants for food caused by the $CO_2$ enrichment, provided better nutrition and therefore resistance to diseases of the day. Disease did not appear to end the rule of the dinosaurs. Currently, the most scientifically accepted hypothesis is that the dino killers were a giant meteor impact and the accompanying large volcanic eruptions.

In summary, a global climate moderately warmer and wetter is certainly easier to tolerate and survive in than a moderately colder, drier, and windier climate. The world's agricultural output certainly would expand, both in more viable, agricultural land at higher latitudes and in growth enhancement because of the presence of more $CO_2$. The historical distribution and abundance of plant and terrestrial animal life appear to support the fact that most life on Earth thrived during warmer, rather than cooler, climates. We have seen that mankind fared much better in the warmer times of the last 2,000 years. Having climate remain right where it is would be mighty fine, but based on history, an absolutely stable climate would be the most unusual climate event that has ever been documented.

While keeping Earth's temperature near its present value might seem ideal, we do not have the technology, in this first decade of the twenty-first century, to be able to "lock in" our current climate. It will change even if we do not tamper with it or if we should get $CO_2$ back down to the preindustrial level of 280 ppm. Still, we have no riskless incentive to push it artificially toward a warmer state or colder state. While we need to study the information in Earth's library to see if it will help us eventually be able to strongly influence some of the climate drivers, we also need to get prepared to deal with whatever climate changes may occur, one way or the other. Same answer, same recommendation: more research.

# Chapter Seven
# GENERALIZATIONS AND SCENARIO DISCUSSIONS

*As the climate cools, global winds get stronger (and drier).*
*As the climate warms, global winds get less strong (and wetter).*

Ice cores bear out these generalizations, but anomalies in the above are commonly thought to be hurricanes and cyclones that derive their energy from the warm surface waters of the oceans. In a general global warming, the average, lighter surface winds could get interrupted by occasional severe storms fueled by these warm surface waters. Recently, however, it has been found that high-velocity, high-altitude winds, during periods of warmer or cooler temperatures, can shear off the tops or tilt over the rotating storms and decrease their intensity. A study by members of Woods Hole Oceanographic Institute that analyzed 5,000 years of sediments in a normally isolated Caribbean lagoon (except during storm surges), has found that the hurricane frequency and intensity were just as high during periods when the temperatures were cooler. In fact, they found that intense hurricanes even made landfall during the latter half of the Little Ice Age. They further found that the times of more and less hurricane activity seemed to run in cycles that were not connected to temperatures in the Atlantic region. So, we must be careful about jumping to conclusions without examining all available data. Results of this study that were reported in Space and Science, 2007, point once again to the complexity of the climate system.

As a couple of scenarios are played out below, keep in mind that they are just that – scenarios. Although scenarios are simply possible outcomes under varying conditions, they help us do broader analyses of what the future might hold, as well as help us understand what might have happened in the past.

The current warming being experienced on Earth will not necessarily lead to a long-term, severe hothouse, but depending on the interaction of the many drivers or dominance by a key one, it could lead to such an outcome. For instance, as Earth begins to heat, the average wind velocity should drop, which would mean fewer nutrients would be blown from land to the open oceans. Diminishing the food available to life in the oceans that takes $CO_2$ from the atmosphere to make shells or skeletons would mean that less $CO_2$ is taken from the atmosphere and sequestered at the bottom of the ocean;

thus allowing a build-up of $CO_2$. Also, lighter winds would cause less mechanical upwelling of nutrient-rich waters that normally stimulate coastal organic blooms that result in removal of $CO_2$ from the atmosphere. This slower removal of $CO_2$ could mean more $CO_2$ could build up in the atmosphere and add some level of warming (positive feedback). Note here that I am saying "could" a lot and not "will" because there are other, more powerful drivers constantly at work. As the surface of the oceans warms, more $CO_2$ would be released to the atmosphere, increasing the greenhouse effect to some degree.

As $CO_2$ increased and the vegetation responded with accelerated growth, more methane would be created from the rotting plants, causing the addition of another greenhouse gas. Also, possibly of even more importance, adding vegetation to cover former highly-reflective, barren land would lower Earth's albedo and result in more solar heat being absorbed at Earth's surface. With continued warming and rising sea levels caused by thermal expansion of the oceans and melting glaciers, sandy coastal plains would get covered with salt water of lower albedo, and covering former permafrost coasts with water also would mean lower albedo. Melting of some of the permafrost would release more methane and could result in more positive feedback for warming.

Polar and mountain glacial melting would mean that more ice-free low-albedo land would be exposed, which could mean still more positive feedback for warming. However, exposing previously buried rocks and soil would cause more chemical weathering and long- term removal of $CO_2$. But as the higher $CO_2$ levels that had been created in the short term stimulated robust plant growth, the lower albedo caused by the greening of land should overpower the much longer-term effect of chemically removing $CO_2$ from the atmosphere.

In the scenario above, with so many feedbacks apparently lined up to favor still more warming, what could possibly happen to save us from such a hellish future? Let's grab some facts, add some speculation, and play out an opposite scenario.

That the orbital alignment is gradually moving toward a cooling effect has not been disputed as far as I know. Some say the man-made $CO_2$ increase may overpower the orbital factor, but I personally doubt it, since $CO_2$'s ability to trap more heat is probably near its effective maximum regarding temperature levels. So, let's continue. The current warming should provide an atmosphere that can, per the laws of physics, hold more water vapor. A higher temperature should cause more evaporation, which can feed the warmer atmosphere with more water vapor. A high level of water vapor (or moisture) in the air generally aids the development of more low-level clouds which block (reflect) some of the sun's rays and also lead to more precipitation, both of which promote cooling. However, since water vapor is by far the most dominant greenhouse gas, the argument can be made that the addition of more water vapor will overpower the cooling effects and lead to more warming. The answer to this question is not yet resolved. But to continue our scenario discussion, let's assume, as many scientists believe, that the cooling influence will win.

As Earth begins to cool, many of the drivers begin to add positive feedback for more cooling such as ice and snow and lower sea levels for increased albedo. Less vegetation in a cooler climate would mean less methane release if the shocking recent study is confirmed that LIVE plants may be releasing 10 percent to 40 percent of the greenhouse gas, methane, into the atmosphere. Even if there is simply less vegetation, there would be less to rot in the water where the methane is routinely formed and more precipitation generally means more chemical weathering, and that would remove $CO_2$ from the atmosphere at a higher rate. Less atmospheric $CO_2$ would mean less $CO_2$ to be utilized by the plants and thus fewer plants covering Earth's surface which would result in a higher albedo.

The suddenly cooler oceans would become a "sink" for removing large quantities of $CO_2$ from the atmosphere, which should cause less greenhouse effect and cause further cooling. The volumetric shrinking through simply cooling the oceans would help lower sea level by a small amount which would increase Earth's albedo somewhat by exposing more high-albedo shorelines.

On top of all of the above scenarios lies a factor that cannot be reliably predicted for the long term. That factor is the total amount of heat and magnetically generated activity and effects coming from the sun. The sun is Earth's ultimate heat engine. The sun's intensity does vary. Proof of that lies in the records of several hundred years of very low sunspot and magnetic activity, called the Maunder Minimum, which, as you will remember, was coincident with the Little Ice Age. Having previously reviewed both real and potential solar effects, you now can grab your pencil and play out several additional warmer-climate or colder-climate scenarios.

It fascinates me how much time and fanfare we give to the greenhouse gas $CO_2$ and how little recognition is given to variations within the sun. In fact, many of the global-warming modelers do not even include solar variations or include test of things such as the cloud-generating potential of cosmic rays in their models. While we cannot do anything about solar variations, we at least can acknowledge the possible contributions of the sun's effects when trying to explain historical global climate changes. We also should pay more attention to the forecasts of future solar cycles so that we might better prepare for solar-generated or solar-enhancing effects on the climate.

You now know the key factors that have been recognized to drive or affect the climate. Have some fun, and make your own predictions. But be careful not to make your prediction by considering the effect of changing only one of the climate drivers and thinking all of the other drivers will temporarily cease to exert their influences.

## Chapter Eight
# THE FUTURE

As you build your own vision of the future or simply follow the researchers' predictions, there will be flaws, certainly in detail. But as research continues and new data flood in to fill the gaps in the summaries gathered here, better-thought-out predictions should emerge that are closer to reality. Regard the source of the new data. Beware of summaries by the media that are designed to sell fear because the media knows alarming news sells better, even in a climate of paradise (shame on them!). Be aware of people who must squeak to get "grease" for their cause; it is human nature to do so. That is an area where the peer review boards should step up and identify some of the purposefully biased reports that appear in technical journals written by people on either extreme of the global-warming issue.

In making your prediction of the future, consider that sea level was nearly 20 feet higher at the end of the last interglacial cycle about 100,000 year ago. It got there without human's input of greenhouse gases. The temperature was warmer than our current temperature. Our sea level could someday rise that much, even more. But it will not rise that much in just a century if it approaches that height at all.

Our current interval of general interglacial warmth may possibly last a few thousand more years because of natural, well-established forces. The orbital alignments that brought us out of the last ice age change very slowly, and this current period of warmth may have some residual orbital influences regarding continued warming. Humanity is changing the atmospheric mix of greenhouse gases that affect climate. Although of a debatable magnitude, most scientists believe the influence has been in enhancing overall warming but many now believe the human-induced increase in greenhouse gas to cause warming is negligible.

Regarding future cooling, we do not know at what point the slowly changing orbital influences will take over and start moving Earth in a cooling direction. Recent temperature measurements by NASA and the Hadley Climate Forecasting Center in the United Kingdom indicate that since year-end 2000, Earth's temperature has "stabilized" from its rather constant recent ascent. Various verbal and written reports from Argentina, Australia, and China have described cooling trends for the last three to seven years. Does this indicate a new trend toward temporary stability or toward cooling? Not necessarily, but it certainly does not support the hypothesis of constant warming caused by today's persistent annual increases in atmospheric $CO_2$ content which are present in Earth's atmosphere 24 hours a day.

There are some scientists now saying that we can predict the near-term stronger and weaker solar cycles. One current prediction by NASA, 2006, is that the solar sunspots and magnetic activity will be extremely weak around 2022 (see Figure 33). This would allow more cosmic rays to reach Earth's lower atmosphere and possibly create more clouds and cool Earth or it could simply mean less radiation and solar wind coming from the sun. How nice to have a prediction that is for a time less than 15 years away, when the forecaster may still be alive and can be held accountable or congratulated for the prediction. With satellites now in place to constantly monitor the sun, predictions of future solar influence should improve.

*Fig. 33*

## THE SUN'S CONVEYOR BELT

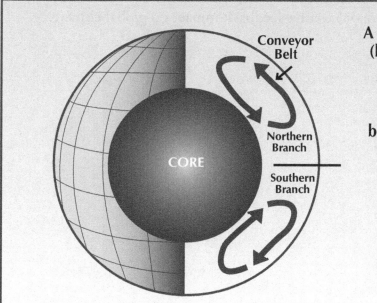

Conveyor Belt

Northern Branch

CORE

Southern Branch

**A massive current of fire (hot plasma) circulates within the Sun.**

**North and South branches take ~40 years to complete a circuit.**

**Average speed since the late 19th century has been 1 m/s, but in recent years the belts have decelerated to 0.75 and 0.35 m/s.**

**The slower belt speeds foretells low solar activity ~20 years into the future.**

**NASA has never seen speeds so slow. "The slowdown we see now should mean the next cycle that will peak around 2022 should be one of the weakest in centuries," per David Hathaway of NASA. Researchers believe the belts control the sunspot cycles. Low sunspot numbers have generally been associated with cooler times on Earth.**

*Adapted after Science@NASA, Long Range Solar Forecast, May 10, 2006*

With confidence I can say the climate will not stay the same. It will continue to change. When you look at any of the examples of temperature curves in this book or in anybody else's book or any research article, you won't see any flat lines! The temperature is always either increasing or decreasing. One day, within no more than a few thousand years, Earth will enter another glacial cycle unless researchers can develop the powerful technology required to overcome the tremendous natural forces that have caused Earth's climate changes for the last several million years.

The good news is that with most of these scenarios, the human race can make a lot of adjustments in a few thousand years and avoid suffering the catastrophic consequences depicted in the media.

If you are prone to worry (some seem to enjoy it), go back and look at the many examples of temperature downspikes that have occurred over very short periods of time such as those that are documented in the Greenland ice cores. Of course, these mainly occurred as Earth was entering, or in the grip of, an ice age. While the downspikes were usually followed by rapid upspikes, I think I know which spikes our ancestors preferred (see Figure 30, page 55)!

Stay tuned. There are many unknowns, and problems to be researched and solved. A prime example is found in the ongoing arguments about something as pervasive and important as water vapor and whether it has a net positive or negative feedback impact on global climate.

# Chapter Nine
# CONCLUSION AND AUTHOR'S OPINIONS

Earth is dynamic, not passive; witness volcanoes and earthquakes, continuously shifting continents, sea-floor spreading centers, shifting ocean currents, changing sea levels, and ever-evolving orbital alignments - always in a state of change. Because the major climate drivers operate on many different timescales it is easy to see why constant change is to be expected. The changes can and have occurred in both directions and at small and large magnitudes and at fast or slow pace. To see just how frequently the changes have occurred over different time intervals look at Figure 34. Since there are no flat lines on any timescale, it does support the fact that Earth's climate is dynamic and never in a state of equilibrium.

Even if we do not add one more molecule of greenhouse gas (water vapor, carbon dioxide, methane, nitrous oxide or ozone) to the atmosphere, the climate will still change! And these climate changes, whether natural or caused by humans, could very well send us in a direction that is not necessarily ideal for the future of humanity. Today, this is the reality, and there is not a whole lot we can do about it. In the meantime, we should not perturb the natural forces unless it is in a direction we are confident will be beneficial.

Should we continue research on how best to capture the $CO_2$ we create and then store it for possible future use? I have already voted 'yes' on this. It is part of the "Don't perturb the system" argument, and it addresses the "Warmer is uncomfortable but colder is catastrophic" reality.

Earth has accumulated a vast supply of fossil fuels such as oil, gas, and coal. The use of these concentrated sources of energy has brought much of humanity to a very high standard of living, including better nutrition and health. However, with more than six billion people on Earth, expected to increase to nine billion in just forty years, the supply of this relatively cheap, efficient energy is being deleted at a rapid rate.

I do believe strongly in developing more renewable energy. But I mean developing it at a reasonable pace, not rushing blindly in and upsetting our overall economy and ability to feed, clothe, shelter and provide medical care for ourselves and much of the rest of the world. Today, five billion of the six and a half billion people who live on Earth are living in somewhat primitive conditions. Electricity, adequate nutrition and proper medical care are critically needed. I don't want us to harm inadvertently a portion of the world's population by taking money away from these other areas in need of research and assistance by making draconian, mandatory reductions in global $CO_2$ emissions.

Many people are clamoring to increase the use of our edible grains, which barely support today's population, as an inefficient, expensive fuel source that has little, if any, positive effect on greenhouse gas reduction. Where is the common sense in this? It does make sense to develop new fuels from renewable sources such as cellulose that do not take food from our tables or food from the animals.

While I believe the world's people deserve a chance to obtain these live-saving necessities of electricity and clean heating, cooling, and cooking power from today's inexpensive energy supplies, I also believe we should vigorously fund and support the development of more efficient appliances and "engines" of all types and the various mechanisms that use the power from these energy "engines".

Let's get the most from our dollars and brain power. Going immediately to higher-cost, less efficient power will strain the world's economies. While the developed countries can afford some of these added costs, any significant reduction in the use of these most efficient and inexpensive energy sources will only ensure that the people in developing nations will not see significant near-term improvements in their lives and well-being. Let's revisit nuclear energy, which has been proved to be safe, affordable, and non-polluting, if we can agree on a suitable storage place for the nuclear waste.

We should support more pure research on how we can best react if natural factors exert a cooling influence on Earth. We may need to "blacken" the surface of Earth to lower Earth's albedo or create more swamps to add more methane to the atmosphere.

What do I believe regarding the issue of global warming? Paleoclimate studies indicate that Earth has warmed somewhat over the last century. These studies also indicate that Earth has cooled over the last 1,000 years relative to the Medieval warm period and then started to warm before any significant change in $CO_2$ content. The climate is in a constant state of change and all of Earth's climate history bears that out. I believe orbital effects, which have been the dominant driver in the past eight out of eight glacial – interglacial cycles, are moving in the direction of pushing the climate of Earth toward another ice age in no more than a few thousand years. $CO_2$ in the atmosphere, although twice its glacial-period level, remains near its historical low (see Figure 4, page 2) and we have found that it follows, not leads, the temperature changes. We also know that the heat-trapping capability of each additional unit of $CO_2$ declines rapidly relative to the capability of the previous unit put into the atmosphere. These last two sea-changing or paradigm-changing revelations are causing a large number of climatologist and paleoclimatologists and scientist in related fields to look at $CO_2$'s effect on climate in a very different light. There may be other such paradigm-changing discoveries that will come from the researchers' laboratories, astronomical observatories, or solar or even extrasolar system space probes.

# NO FLAT LINES!

*Fig. 34*

FROM
OCEAN
CORES

FROM
ICE CORES–
ANTARCTICA

FROM
ICE CORES–
GREENLAND

FROM
THERMOMETER
RECORDS—USA

*Compiled by H. Leighton Steward*

We need to determine, if possible, at what atmospheric saturation level $CO_2$ becomes a nonissue relative to further warming and whether that saturation level has already been reached (see Figure 16, page 23). That said, I am not advocating pouring an unlimited amount of greenhouse gases into the atmosphere because the unlimited addition of the gases might affect in a negative way the chemistry of Earth's oceans or some other as yet unidentified, environmental impactor. More research is needed in the area of $CO_2$'s effect on ocean chemistry and aquatic life's sensitivity to any such changes.

The near-term movement of global climate into a new cold cycle, short of a huge extraterrestrial impact or a super volcanic explosion, could be the most destructive event for the six and a half billion people on Earth that we can imagine. How near? In our or our children's or grandchildren's lifetimes? I'm not predicting exactly when it will occur, because I simply don't know. I'm just saying I believe it will happen within the next several thousand years.

What else can we do? Now that Earth's library is more interpretable, let's accelerate our research of the past so we can better understand what has previously happened in reaction to changes with these individual and collective climate drivers. Then we will better know what to do if confronted with a situation that could negatively affect us or our descendants. Let us borrow advice from Winston Churchill, who reportedly said, "The farther backward you look, the farther forward you are likely to see."

What should we not do? We should not spend "trillions of dollars" on a quick fix for a potential problem that has not been openly debated in a neutral forum by qualified experts. In my opinion, humans are much more likely to vanish from this Earth as a result of a virulent and resistant disease for which we have no cure than we are from either global warming or global cooling. Let's spend the lion's share of the "trillions of dollars" on improving overall living conditions for humanity, and carve out a smaller but significant portion for continued research on global climate change, which is still poorly understood.

We should not move forward based on emotional appeals or feelings that have not stood the test of scientific debate. While many reputable scientists have publicly accepted the earlier consensus that $CO_2$ is the dominant driver in global warming and are now reluctant to say they might have been wrong, many others have stepped forward and admitted that in light of new scientific evidence, they have changed their opinion. Of course, I may be wrong in downplaying $CO_2$'s future warming influence, but I believe this look back at Earth's earlier history and the associated empirical data supports my conclusions. I hope you and many scientists, as well as the legislators that distribute research funding, will take this summary data and consider the implications. Remember some sage advice from an early proponent of common sense; Mark Twain said, "It ain't what you know that gets you into trouble, its what you know you know for sure that just ain't so." Beware of those who profess to know for sure but will not participate in open debates with scientists or others who have a different view or interpretation of the data currently available.

A discussion of global climate change, particularly the current global warming trend, brings on spirited debates about the causes of the warming in private gatherings but, as I have said, I have been very disappointed in the lack of public debate among top scientists in the field of climatology. If I have written of facts or issues that seem controversial, I hope it is a catalyst for more public discussion. Demand the debates and demand that the moderators of the debates be fair and balanced and not trying to steer the conclusions toward their own pre-conceived opinions.

Since all climate drivers are on a different timescale relative to their effect on climate change, I have made a summary chart to remind you which ones we need to worry about today or in the next few decades and which ones your heirs will not need to be concerned with for millions of years. The chart also lists which drivers we can try to influence and which ones we cannot change (see Figure 35).

As for the scenarios of the future, there are plenty to tease your imagination. There are people out there, or even reading this book, that may have the genius to imagine all of the interactions at once and to predict what measures we should take to try to counter the ill effects of the negative scenarios. Maybe it's just a "computer whiz" who finally comes up with the model that considers all significant factors and assigns them the proper weighting and spits out the right recommendations or actions we should take or attempt to ensure a better future

As I said in the Introduction, one of my goals in writing this book was to provide you armchair paleoclimatologists a chance to review some of the summary charts and figures and better understand what is or was happening in all the areas that influence the climate. New or better data are coming in every day. What we do not want to do is move ahead based on faulty data. So, come join us! Go to school. Go back to school. This is no short term exercise. Your children and grandchildren will be the beneficiaries of still better data that will improve on our current observations. Should things change very rapidly, we will have to go with courses of action we can collectively agree to try to preserve our earthly paradise. Let's base any such actions on science, not emotion.

Where and how we spend our money is critically important. Steering where this money goes is primarily decided in Washington, with lesser amounts determined at state levels. Very few of our legislators and administration officials have any education in the natural sciences, much less meteorology. The same is true for the media and also of many of the high-profile intellectuals often chosen to be interviewed. Climate change, and particularly today's global-warming decisions, should not be based on ill-informed, emotional appeals.

Better-informed citizens are important because that is where the votes originate. A lot of pressure can be put on lawmakers and officials who oversee the funding for research. Hence this book, which is intended to give a nonscientific audience and politicians and even judges a general, "soft" scientific review of the most salient factors governing global climate change. We all need to be equipped to come to our own, better-informed conclusions rather than merely being slaves to media manipulation or politically or financially motivated information sources.

*Fig. 35*

| CANNOT CHANGE | HOW SOON MIGHT WE FEEL THE CHANGE |
|---|---|
| Sun's strength | Immediately |
| Orbital eccentricity | Slowly |
| Earth's tilt | Slowly |
| Earth's wobble | Slowly |
| Ocean currents | Decades or centuries |
| Plate tectonics | Millions of years |
| Location of continents | Millions of years |
| Elevation of land masses | Millions of years |
| Volcanism | Immediately |
| Atmospheric circulation | Constantly |
| Water vapor | Constantly |
| Cosmic rays | Immediately |

| CAN CHANGE | |
|---|---|
| GHG $CO_2$ | Decades or centuries |
| GHG methane | Immediately (clathrates) |
| Chemical weathering (slightly) | Very slowly |
| E.T. impacts (divert) | Immediately |
| Fauna and flora | Decade |
| Albedo | Decade |

*H. Leighton Steward, 2007*

Finally, think about how the current hysteria over $CO_2$ - induced global catastrophes is affecting our children and grandchildren. They are constantly presented with frightening climate information, often by ill-informed educators in the schools. It has been said that children today are more afraid of global warming than of terrorists, cancer, or vehicular deaths. Open, vigorous debates by climate experts will bring much broader, more balanced information to the public, politicians, and educators, and through them, to our children. Let's hope everybody including our children can enjoy this current climate paradise by seeing how it compares to other periods of Earth's history.

I promised to give you a bullet point list of the key factors relative to Earth's climate change.

# GLOBAL CLIMATE SUMMARY

- Climate is continuously changing; there are no flat spots on any temperature curves.
- Constant change is expected since the climate drivers operate at different time scales
- Earth has been generally warming for 250 years and the cause is still being argued; the science has not been settled and public debates are needed.
- Technology today enables the reading of Earth's paleoclimate "library" and gives clues as to causes of these earlier climate changes; demand more research
- Changes in levels of $CO_2$ have been determined to follow (lag) temperature changes; a cause does not follow an effect.
- $CO_2$'s ability to trap heat declines rapidly, logarithmically, and quickly reaches a point of significantly reduced future effect. Many scientists say we are already at the point of insignificance. This, the lag effect, and the negative correlations of temperature and $CO_2$ from empirical observations in the not too distant past, denies the "$CO_2$ is the dominant cause of global warming" hypothesis.
- The public's infatuation with $CO_2$ as the primary cause of climate warming is inappropriate but we should still reduce our unprecedented input and sequester as much as is technically reasonable because there may be non-climate, negative effects that are currently not recognized.
- The supply of fossil fuels is limited and we should vigorously pursue improved efficiencies of energy dependent devices and development of reasonable alternative fuels.
- Modelers should be evaluating <u>all</u> the key drivers that affect climate change and researching what mankind might be able to do to temper negative climate changes.

Lastly, reflect on the data presented in *Fire, Ice and Paradise* and then heed another sage quote (see Figure 36). Also included is a discussion on myths that cover my views on many questions you may have pondered that can be found in Chapter 10.

---

**IF THE FACTS CHANGE, I'LL CHANGE MY OPINION.**

**WHAT DO YOU DO, SIR?**

*Fig. 36*

*John Maynard Keynes*

---

## Chapter Ten
# POPULAR MYTHS OR GROSS EXAGGERATIONS

**Myth:** The graph by Mann, et al., the so-called "hockey stick" used in the 2001 Intergovernmental Panel on Climate Change (IPCC) report, showing 1,000 years of stable temperatures until the twentieth century when the temperatures rose rapidly after humans began putting large quantities of $CO_2$ into the atmosphere, proved that $CO_2$ has caused global warming.

**Fact:** Later, the IPCC rejected the validity of the Mann graph, since it left out the data that showed the climate had varied considerably over the past 1,000 years before humans introduced significant amounts of $CO_2$ into the atmosphere (see Figure 37, page 93). Since $CO_2$ stayed level at 280 ppm for these same 1,000 years before the large input of $CO_2$ by industrialization in the mid-1900s, the corrected graph proves the opposite, that $CO_2$ has <u>not</u> been the driver of climate change.

**Myth:** The close correlation of $CO_2$ and temperature, as temperature has gone up and down over the last 400,000 years, proves that $CO_2$ is causing the climate changes (see Figure 12, page 17).

**Fact:** Since 1999, multiple technical, peer reviewed articles have been available that demonstrate exactly the opposite conclusion; that $CO_2$ lagged temperature changes going both up and down. A cause does not follow an effect. Temperatures changed and then, several hundred years later, the $CO_2$ levels changed. Since a cause does not follow an effect, this proves that $CO_2$ is not a primary driver of climate change.

**Myth:** Sea level likely will rise 20 <u>feet</u> by the end of the century.

**Fact:** Even the IPCC says the most likely rise will be 17 <u>inches</u> while most climatologists predict a rise of only 7 or 8 inches (about 3 millimeters a year).

**Myth:** Scientists are unanimous that man-made $CO_2$ is the dominant cause of global warming.

**Fact:** Not so. Many, many reputable scientists believe that natural factors overpower the current influence of $CO_2$ on global warming. Several hundred prominent scientists and/or science professors that have no ties to the petroleum industry have stated publicly that $CO_2$ is not a significant cause of global warming.

**Myth**: Higher temperatures kill people.

**Fact**: A half-truth at best. High temperatures can cause deaths but according to health officials, many more people die each year from cold weather than from hot weather. Just read the news after the strong winter storms.

*Fig. 37*

**TWO PALEOTEMPERATURE INTERPRETATIONS**

A graph appeared in the literature and the Intergovernmental Panel on Climate Change (IPCC), 2001 report (with temperature grossly generalized to within the grey band above), to "prove" man's impact on global warming (the upward temperature rise from the 1970s to 2000), rather than the historically documented graph (red curve), that shows large climate swings prior to any significant contribution of greenhouse gases by man. The climate changed considerably with essentially no change in $CO_2$ content prior to 1900. IPCC subsequently withdrew the earlier interpretation citing improper use of data in its construction.

*Modified after McIntyre, et al., 2005*

**Myth**: Warmer weather will cause more malaria.

**Fact**: According to Dr. Paul Reiter of the world-renowned Pasteur Institute, this is not true (Reiter, Paul, 2000). Malaria was much more prevalent and had already extended into Siberia before the most recent warming.

**Myth**: Global warming will adversely affect worldwide food production.

**Fact**: Exactly the opposite. Warmer weather will open much more land to agriculture (at higher latitudes), and the rising $CO_2$ will, as it already has, stimulate plant growth.

**Myth**: The current rate of warming is unprecedented.

**Fact**: The short-term temperature swings during the most recent glacial stage had much higher rates of cooling and warming. Also, the rate of temperature rise in the 1920s and 1930s, before people began putting huge amounts of $CO_2$ into the atmosphere, was higher than the rate of increase over the last 30 years (see Figure 38).

**Myth**: The United States is the largest contributor of man-made $CO_2$.

**Fact**: China, which has no $CO_2$ restrictions per the Kyoto "agreement," has recently exceeded the United States in $CO_2$ output and the gap is growing very rapidly because China is opening a new coal-fired power plant every WEEK and its production of automobiles is growing at a much more rapid pace than that occurring in the United States.

**Myth**: Storms are now much more frequent and intense because of global warming.

**Fact**: According to the National Hurricane Center, storms are no more intense or frequent than they were in the 1930s when temperatures in the United States were as high as current temperatures but when there was much less $CO_2$ in the atmosphere. Constant, 24/7 media coverage of every significant storm worldwide just makes it seem that way (see Figure 39). Insist on the facts; not just what some individual or reporter says to support their cause.

**Myth**: Large chunks of ice breaking off glaciers (calving) around the world are proof of the catastrophe being caused by global warming.

**Fact**: Glaciers constantly calve ice into the sea even when the glaciers are growing. Photographs of such activity are no proof of any catastrophe occurring.

**Myth**: Human activities are causing unprecedented melting of Earth's glaciers and the accompanying sea level rise.

**Fact**: Ask yourself, why did so many more glaciers melt and sea level rise 18 to 20 feet higher than today during the last interglacial period, 115,000 years ago, when modern man was not around to introduce any $CO_2$ into the atmosphere? Ponder the facts.

*Fig. 38*

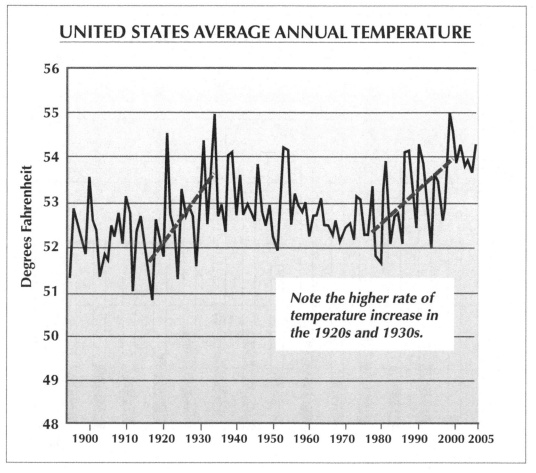

**UNITED STATES AVERAGE ANNUAL TEMPERATURE**

*Note the higher rate of temperature increase in the 1920s and 1930s.*

**Our climate roller coaster continues.**

*SOURCE: National Climatic Data Center*
*Modified after Christopher C. Horner, 2007*

**Myth**: As Earth warms, the climate will become much drier and windier.

**Fact**: All studies of the ice cores prove just the opposite. The colder times were both windier and drier.

**Myth**: For every additional unit of $CO_2$ that we add to the atmosphere, we will see an equal amount of warming.

**Fact**: False. Textbooks on thermodynamics and calculations by many reputable scientists show that each additional unit of $CO_2$ will trap less heat than the previous unit. The decline is rapid, logarithmic, causing $CO_2$'s warming effect to become minimized very quickly. Since the current pace of the temperature rise has been less than the annual $CO_2$ rise, it supports the premise of $CO_2$'s diminishing role in causing climate change. In fact, Earth's temperature has essentially "stabilized" for the last seven years as worldwide $CO_2$ levels have continued to rise.

Fig. 39

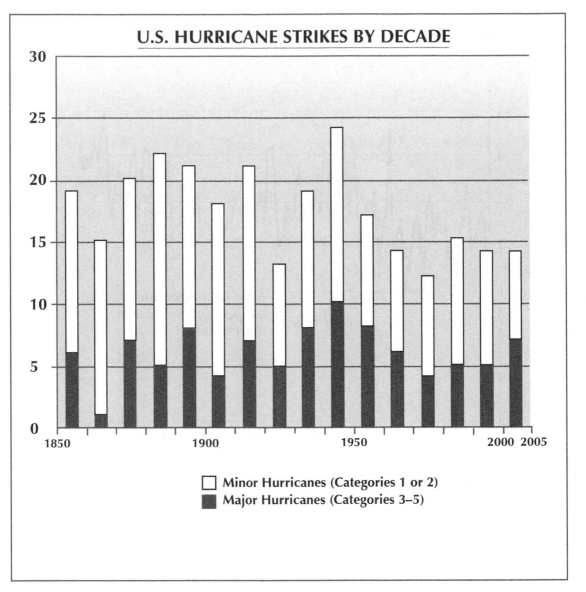

**U.S. HURRICANE STRIKES BY DECADE**

Minor Hurricanes (Categories 1 or 2)
Major Hurricanes (Categories 3–5)

SOURCE: National Climatic Data Center
Modified after Christopher C. Horner, 2007

# RECOMMENDED READINGS

Summarized below are a few books that address one phase or another of global climate change. All are relatively informative, and several are very well written. One textbook is included for those of you who want to know what you would learn if you took an intensive college-level course in this subject or, better yet, pursued a major in this field. These books are referenced in the bibliography, even though they may not be specifically referenced in the text. How, for instance, could a researcher ever figure out who first coined the term "driver" or "indicator", or spoke of $CO_2$ being an important greenhouse gas or that climate changes have affected past civilizations? If I knew, I would certainly reference them. Two important factors are missing from nearly all of these books that mention global warming. One is that the $CO_2$ levels follow temperature changes, and the other is that $CO_2$ quickly loses its ability to trap more heat, facts that may not have been known at the time some of their manuscripts were in production.

*A Brain for All Seasons* by William H. Calvin, M.D. (2002). Dr. Calvin, a professor of psychiatry, writes a great summation on global climate change and how the constant stress caused by the changes resulted in human brains either evolving to cope with the change or becoming another dead-end species or line of individuals.

*The Eternal Frontier* by Tim Flannery (2001). This is a paleohistory of North America from plants to animals to early humans. It is very well written, and it should cause you to look at this continent in a way that will make you want to go sit on a Wyoming mountaintop and dream of what North America looked like all those years prior to our ancestor's recent arrival.

*The Two Mile Time Machine* by Professor of Geosciences, Richard B. Alley (2000). Dr. Alley was directly involved in coring and interpreting a two mile long ice core taken from the Greenland ice sheet. His analyses helped alert the world that many global climate changes have occurred in periods as short as a few years to a few decades.

*The Ice Chronicles* by Paul A. Mayewski and Frank White (2002). Paul Mayewski was the chief scientist on the expedition that drilled the two-mile ice core mentioned above. Frank White is a writer. Mayewski believes global warming is caused by humans but advocates a fairly balanced approach to greenhouse gas reduction accompanied by more research.

***Geological Perspectives of Global Climate Change*** by Lee C. Gerhard, et al. (2001). This is a compendium of chapters by various scientists that addresses multiple causes and effects of climate change and demonstrates that the climate has changed continuously before any influence by humans. It also contains a chapter by Sherwood Idso that summarizes 20 years of real world research designed to determine Earth's sensitivity to changes in the levels of greenhouse gases.

***Out of Thin Air*** by Peter D. Ward, (2006). Ward presents a very different viewpoint on the evolution of plant and animal life in that he postulates that the changing levels of atmospheric oxygen were the driving mechanism for evolutionary change and that those species that adapted well, such as the dinosaurs, became the dominant species of their time.

***Earth's Climate, Past and Future*** by William F. Ruddiman (2000). This is a college-level textbook and the best written and illustrated of any of the several I have read. Ruddiman's coverage of global climate is broad, well documented, and quite readable and understandable for someone with a modest background in science.

# ACKNOWLEDGEMENTS

An acknowledgement is often taken as an indication that the people who have encouraged or assisted the author tacitly agree with the conclusions expressed in the author's book. The conclusions reached in this book are strictly my own. I have no problem with anyone who has come to a different interpretation of any of my conclusions, and I encourage you, the reader, to utilize the scientific data presented in this book, plus any other scientific data you may find, to arrive at your own conclusions. But do not take the statements as meaning I have unexpressed concerns about my conclusions or recommendations, because until new scientific data come in to change my mind, these are my conclusions and recommendations and I'm sticking with them!

The first acknowledgment is to Dr. Crayton Yapp, professor of geological sciences at Southern Methodist University (SMU), who is an expert in the use of stable isotopes to determine the paleoclimate history of Earth. Crayton's telling me of $CO_2$ levels in the atmosphere hundreds of millions of years ago that were ten or more times higher than in the atmosphere today got me interested in this subject. After a cursory look at how the temperature of Earth, as well as the concentrations of greenhouse gases, have moved up and down over the eons, I thought it would be useful to try to summarize those facts for the general, nonscientific public as well as for our policymakers, few of whom have any science background. This leaves most people at the mercy of the cryptic summaries provided by the media or by constituents who may have their own interests in mind.

After they saw a crude outline of my proposed book (no conclusions yet), I received encouragement, questions, and advice from Dr. Jim Brooks, one of my geology professors, and Dr. Louis Jacobs. These two very fine gentlemen are vice chairman and president, respectively, of the Institute for the Study of Earth and Man at SMU, an organization that has a mission of bringing mankind and the environment together through the better understanding of geology, archeology and anthropology. I received further assistance from Dr. Bob Gregory, chairman of the department of geological sciences at SMU, Professor Crayton Yapp and Associate Professors Bonnie Jacobs and Neil Tabor. Several spirited debates ensued!

Diana Vineyard, a PhD candidate about ready for prime time, helped me find many of the scientific articles that were important to my research.

On the domestic front, my wife of several decades, Lynda, who was a fashion design major, read an early draft and bounded in smiling. I took that to mean that a nonscientist might be able to grasp the essence of global climate change. Lynda named my first book, (Sugar Busters ®) which with its offspring has sold more than four million copies. I know, I know, this book has no such chance, because it will not help you lose weight and improve your blood chemistry, but it may also change the way you and your descendants live in the future. Lynda has had to put up with my stacks and boxes of research papers piled on her dining room table, kitchen table, and bar for more than two years. Her early enthusiasm for my writing of *Fire, Ice and Paradise* did not waiver even when she realized she had lost me, not for three or four months, but for more than two years.

I received encouragement from several of my old geology friends; Gene Shinn, Gene Ames, Joel Wilkinson, Jim Nielson, Jim Gibbs, Dr. Pete Rose, Bob Walpole and Dave Haglund, who saw early versions of the text or the illustrations. Dr. Gene Shinn and Gene Ames were particularly helpful by furnishing me scientific reports I might otherwise have overlooked. I received some very good information from Drs. Lee Gerhard and Ray Thomasson. Lee has been a lead author and lecturer on global climate change and Ray probably will write a more technical book on this extremely complex subject. None of these fine folks believes that the scientific community has come to a consensus on what is the major cause of global warming or global climate change.

Jan Woods-Krier professionalized my illustrations and patiently put up with my many tweaks and changes. Maurine Maloy gave me prompt turnaround on the early and final drafts of my manuscript. My editor for **Sugar Busters!**, Maureen O'Neal, who had learned to read my scribbling previously, gave me both editorial and common sense advice.

In writing a non-fiction book that includes some technical illustrations, I discovered an "eternity" factor. It takes an eternity to get permissions to use some of the illustrations. I would still be waiting if not for the advice and help of another fine lady, Mary Kay Grosvald, librarian for the AAPG in Tulsa, Oklahoma.

Lastly, I thank everyone who saw the need and then encouraged me to write a book on the soft science of this subject written at the layman, legislative and administrative level.

Thank you all!

# GLOSSARY

**Aerosols** – very small particles or droplets suspended in the air. They are conducive to forming water droplets and clouds.

**Albedo** – the percentage of incoming radiation from the sun that is reflected away from the Earth's surface or atmosphere.

**Amphibian** – a form of animal life that spends part of the time in the water and the other part on land.

**Anthropogenic** – caused by human activities.

**AGW** – anthropogenic (human-caused) global warming.

**Aquatic** – growing or living in or upon the water.

**Axial precession** – the wobbling of Earth's rotation (spinning) that causes Earth's axis of rotation (aligned from north to south pole) to point in different directions over cycles of 19,000 to 23,000 years.

**Carbon dioxide** – a greenhouse gas made of one part carbon and two parts oxygen ($CO_2$).

**Carbonate rocks** – rocks like limestones made primarily of one part calcium, one part carbon and three parts oxygen ($CaCO_3$). Some contain magnesium (dolomites).

**Celsius** – a temperature scale on which water at sea level freezes at zero degrees and boils at 100 degrees.

**Chemical weathering** – alteration or dissolving of minerals to a different form when exposed to contact with water and the atmosphere.

**Clathrate** – a form of matter in which a substance such as methane resides in a frozen state.

**Comet** – a frozen mass of dust and gas orbiting the sun. Comets occasionally collide with Earth or Earth's atmosphere.

**Continental crust** – the outermost layer of Earth's surface that averages approximately 20 miles (35 kilometers) thick beneath the continents.

**Continental drift** – the lateral movements of rigid (lithospheric) plates that lie above the molten mantle layer within Earth that occurs at approximately 60 miles (100 kilometers) below Earth's surface.

**Convection cells** – motions within the molten layers of Earth where hotter molten areas rise and cooler areas sink, forming circular loops that cause the rigid plate above to move or drift.

**Cosmic rays** – high-energy particles from outer space that bombard Earth and Earth's atmosphere and produce secondary microparticles in Earth's lower atmosphere, some of which are believed to help form low level clouds that cool Earth.

**Crustal plates** – the outer layers of Earth that average about 3.5 miles (6 kilometers) thick beneath the oceans and 20 miles (35 kilometers) thick beneath the continents.

**Drivers** – used herein as a cause or reinforcement of climate change, including global warming.

**El Niño** – a climatic pattern that develops in the eastern tropical Pacific every two to seven years that is marked by warmer-than-average surface waters that produce a significant regional, if not global, climate effect.

**Elliptical orbit** – not round but more oval in shape and not centered equally around the object being orbited (in this case the sun).

**Evaporites** – rocks or minerals that form by precipitation of crystals from water evaporating in restricted basins in arid climates.

**Faculae** – bright rings or halos that surround the dark sunspots on the surface of the sun. While the sunspots themselves produce a cooler-than-average surface, the extra energy produced by the faculae more than makes up for the cooler sunspots.

**Fahrenheit** – a temperature scale on which water freezes at 32 degrees and boils at 212 degrees.

**Fauna** – the animals of Earth.

**Feedback** – a process in Earth's climate system that amplifies or moderates changes occurring in the climate.

**Flora** – the plants of Earth.

**Gondwana** – an early, pre-300 mya supercontinent in the Southern Hemisphere.

**Greenhouse effect** – the warming of Earth's surface or atmosphere by gases such as water vapor, $CO_2$, or methane temporarily absorbing some of the long-wave, infrared heat radiated from Earth's surface that helps keep Earth's near-surface temperature above freezing.

**Greenhouse gas** – an atmospheric gas that temporarily absorbs and radiates infrared energy (heat) and helps keep Earth's near-surface temperature above freezing.

**Half-life** – the time it takes for one half of the number of atoms of a radioactive isotope to decay.

**Hothouse** – an extended period of time in which Earth has been, or will be, unusually warm.

**Hydrocarbon** – an organic compound containing only hydrogen and carbon. Three of the currently most efficient and commonly used energy sources are the hydrocarbons oil, natural gas, and coal.

**Icehouse** – an extended period of time in which Earth has been or will be unusually cold, e.g. a glacial period.

**Indicators** – used herein to describe things that give us clues or direct measurements of the older (paleo) climate's temperature and atmospheric composition or energy.

**Infrared** – radiation made up of nonvisible "light" of longer wavelengths. The heat energy radiated from Earth's surface is infrared, nonvisible radiation.

**Insolation** – the amount of the sun's energy (radiation) arriving at the top of Earth's atmosphere.

**Ionize** – to change or be changed into electrically charged ions, some of which may help in the formation of water droplets that form rain or clouds.

**Irradiance** – the amount of light or other radiant energy striking a specific surface area.

**Isotope** – form of a chemical element that has the same number of protons but a different number of neutrons in the nucleus, which gives the element a different atomic weight.

**La Niña** – a climate pattern that develops in the eastern tropical Pacific that is marked by cooler-than-average surface waters that produce a significant regional, if not global, climate effect.

**Laurasia** – a very large continent that formed in the Northern Hemisphere before approximately 300 mya.

**Little Ice Age** – a period of time from approximately the 1300s to the 1800s, when Earth was colder than the immediately preceding or succeeding periods. The coldest part of the Little Ice Age was coincident with a period of very low sunspot activity that lasted from 1645 A.D. to 1715 A.D.

**Luminosity** – brightness or the giving off of light.

**Maunder Minimum** – a time of very low sunspot activity that lasted from 1645 to 1715 A.D. and coincided with (if not caused) the coldest part of the Little Ice Age.

**Mechanical weathering** – the physical breaking up, exposing, or washing away, or blowing away of rocks or sediments.

**Medieval Warm Period or Climate Optimum** – a period around 1100 to 1300 A.D. when the Northern Hemisphere was one or two degrees celsius warmer than today.

**Meteor** – a solid body of rock or metal from outer space that is orbiting or traveling within the solar system.

**Meteorite** – a solid body of rock or metal that comes from outer space or the outer solar system and strikes Earth's surface; the surviving pieces are called meteorites.

**Mid-ocean spreading centers** – the place or elongated ridge where two of Earth's plates are moving apart, with the ensuing void being constantly filled with molten rock from deeper within Earth.

**Millennium** – one thousand years in length, or describing a major period that contains 1,000 years, such as 1,000 A.D. to 2,000 A.D.

**Negative feedback** – a process that moderates or counters the original perturbation of Earth's climate.

**Nitros oxide** – a colorless, nonflammable greenhouse gas made up of two parts nitrogen and one part oxygen ($N_2O$).

**Oceanic crust** – the outermost layer of Earth's surface, lying beneath the oceans, that averages approximately 3.5 miles (6 kilometers) thick.

**Outcrop** – the exposures at Earth's surface where layers or deposits of rock or sediment can be seen and accessed.

**Ozone** – a triple molecule of oxygen ($O_3$) formed by the collision of cosmic particles and ordinary oxygen. Ozone is a weak greenhouse gas that also helps block ultraviolet radiation which can cause cancer.

**Pangaea** – the supercontinent that was formed when all of Earth's continents coalesced into one giant continent, after about 300 mya.

**Particulates** – very small particles of various origins that can help form water droplets, which can precipitate or form clouds.

**Photosynthesis** – the process by which plants absorb sunlight and use it to convert water and $CO_2$ to carbohydrates (plant tissue) and release oxygen in the process.

**Plate tectonics** – effects caused by the lateral motions of the rigid outermost layers of Earth. These individual or collective plates collide and cause mountain building, slide past each other, or tear apart (spread).

**Pollen** – the soft fertilizing elements of a plant, the more rigid walls of which can sometimes survive burial and transformation into solid rock.

**Positive feedback** – a process that adds to the original perturbation of Earth's climate.

**Precession** – a wobbling effect of a spinning body such as a top (or the Earth) when the torque changes the position of its axis of rotation, generally changing the axis in the shape of a circular cone as it returns to its original position.

**Preindustrial** – refers herein to the time in the early part of the 1900s before the $CO_2$ content began to rise significantly due to man's accelerated industrial activities.

**Radiation** – ultraviolet and visible photons sent from the sun to Earth.

**Roman Warm Period** – a period of above average global temperatures that peaked around 2,000 years ago which saw the Alpine glaciers retreat nearly 1,000 feet higher than at present.

**Seafloor spreading centers** – the same as mid-ocean spreading center.

**Sequester** – store or put away.

**Silicious rocks** – rocks that contain one part silica and two parts oxygen. These rocks, when chemically weathered, react and take $CO_2$ from the atmosphere.

**Snowball Earth** – a time or times between approximately 850 and 650 mya when some glacial evidence is found on all continents, even those near the equator, suggesting the entire surface of Earth may have been frozen.

**Solar system** – the sun and all of the planets or smaller bodies that orbit the sun.

**Stratosphere** – the atmosphere that extends from the top of the troposphere, about 6 miles, up to about 30 miles above Earth's surface.

**Subduction zone** – the area where two of Earth's rigid plates collide and one plate gets shoved under, or subducted, beneath the other.

**Sunspot cycle** – a natural 11-year cycle of a varying number of visible sunspots on the surface of the sun. The cycles can and do vary in length and intensity.

**Sunspot cycle length** – the actual length of a particular sunspot cycle which normally is 11 years.

**Tilt axis** – the angle between Earth's equatorial plane and the plane of Earth's orbit around the sun. The angle changes from about 21.8 to 24.4 degrees over a 41,000-year time period.

**Troposphere** – the lower part of Earth's atmosphere that extends up to as much as 35,000 feet (7 miles or 10 kilometers) above Earth's surface.

**Ultraviolet light** – invisible, short-wavelength light. This is the type of light that comes from the sun.

**Upwelling** – the rising of deep, cool, and generally nutrient-rich waters to the ocean's surface.

**Water vapor** – water that is in a gaseous state in the atmosphere. When the water vapor condenses into droplets, it becomes either clouds, rain, or ice crystals.

# BIBLIOGRAPHY

Abbott, Dallas, 2006, *NY Times, Science Times*, Nov. 14, 2006.

Abbott, Dallas, 2008, professor and adjunct research scientist at Lamont-Doherty Earth Observatory at Columbia University, Palisades, New York, personal communication.

Alley, R.B., 2000, *The Two Mile Time Machine*, Princeton, New Jersey, Princeton University Press.

Ambrose, Stanley H., 1998, "Late Pleistocene Human Population Bottlenecks", *J. of Human Evolution* 34 (6), p 623-51.

Anderson & Borns, 1997, *AAPG Studies in Geology #47*, AAPG, Tulsa, Ok., p 195.

Archibald, David, 2007, Personal Communication

Archibald, David, 2007, Lavorsier Conference Presentation, Melbourne, Australia, http://www.lavorsier.com.au/papers/Conf2007/Hammer2007.pdf.

Berner, R. A., 2006, "Geocarb-sulf: A combined Model for Phaneroazoic $O_2$ and $CO_2$", *Geochem. Cosmochim.* Acta 70: 5653-64.

Berner, R.A., 2004, *The Phanerozoic Carbon Cycle: $CO_2$ and $O_2$*, New York: Oxford University Press.

Berner, R. A., 1994, "Geocarb III, A revised Model of $CO_2$", *Am. J of Science*, v. 291: p 56-91.

Bluemle, J. P., et al, 2001, "Rate & Magnitude of Past Global Climate Changes", *AAPG Studies in Geol. #47*: Tulsa, Ok, p 193-212.

Bond, Gerald, et al., 2001, "Persistent Solar Influence", *Science*, v. 294: p 2130-36.

Bond, Gerald, et al., 1997, "A Persuasive Millennial Cycle", *Science*, v. 278: p 1257-66.

Broecker, W. S., 1991, "The Great Ocean Conveyor", *Oceanography*, v. 4: p 79-89.

Broecker, W. S., 1992, "Thermohaline Circulation", *Science*, v. 278: p 1582-88.

Broecker, W. S., 2001, "Was The Medieval Warm Period Global?", *Science*, v. 291: p 1497-99.

Callion, N., et al., 2003, "Timing of Atmospheric $CO_2$ and Antarctic Temperature Changes Across Termination III", *Science*, v. 299: p 1728-31.

Calvin, W. H., 2002, *A Brain for All Seasons*, Chicago: University of Chicago Press.

Crowley, T. J. and North, G. R., 1991, *Paleoclimatology*, New York: Oxford University Press.

Cuffey, K.M., et al., 1994, "Calibration of the $^{18}O$ Isotopic Paleothermometer", *J of Glacialology*, v. 40: p 341-49.

Daly, S.I., 2000, "The Hockey Stick": www.john-daly.com/hockey/hockey.htm.

Dansgaard, W., et al., 1993, "Evidence for Instability of Past Climate", *Nature*, v. 364: p 218-20.

Fans, S. M., et al., 1998, "A Large North America Carbon Sink", *Science*, v. 282: p 442-446.

Fischer, H. M., et al., 1999, "Ice Core Records of Atmospheric $CO_2$", *Science*, v. 283: p 1712-14.

Flannery, Tim, 2001, *The Eternal Frontier*, New York: Atlantic Monthly Press.

Frakes, L.A., 2005, *Climate Modes of the Phanerozoic*, Cambridge, U.K.: Cambridge University Press.

Friis – Christensen, E. & Lassen, K., 1991, "Length of the Solar Cycle, An Indicator of Solar Activity", *Science*, v. 254: p 698-700.

Gerhard, Lee C., 2001, "Climate Change, Conflict of Observational Science, Theory, and Politics", *AAPG Bulletin*: v. 88, p 1211-20.

Gerhard, Lee C., et al., 2001, "Geological Perspectives of Global Climate Change", *AAPG Studies in Geology #47*: AAPG, Tulsa, Ok.

Gerhard, Lee C., 2007, *Search and Discovery*, www.searchanddiscovery.net//documents/2007/07005gerhard/index.htm.

Gerhard, Lee C., 2007, former principal geologist of the Kansas Geological Survey and State Geologist, personal communication.

Gordon, W. A., 1975, *Journal of Geology*, v. 83: p 671-84.

Gore, Al, 2006, *An Inconvenient Truth*, New York, Rodale Books.

Gradstein, F.M., et al., 2004, *A Geologic Time Scale*, Cambridge, U.K.: Cambridge University Press.

Gregory, Robert, 2007, Chmn. Dept. of Geological Sciences, Southern Methodist University: personal communication.

Guslakov, V. K., 2006, "Ancient Crash, Epic Wave", *New York Times, Science Times*: Nov. 14, 2006.

Hallam A., 1984, "Pre-Quaternary Sea Level Changes", *Earth Planetary Science* 12: 205-43.

Hieb, M. 2003, "Global warming: A Closer Look at the Numbers", in Hieb, M. and Hieb, H., 2003, "Global Warming, A Chilling Perspective": www.clearlight.com/~mhieb/WVFossils/Ice_Ages.html.

Horner, Christopher, 2007, *The Politically Incorrect Guide to Global Warming*, Washington, D.C.: Regnery Publishing, 350 pgs and personal communication.

Hoyt, D.V., 2006, Personal Communication to Gerhard, Lee C.

Hoyt, D. V. and Schatlen, K. H., 1997, *The Role of the Sun in Climate Change*, New York: Oxford Univ. Press, 279 p.

Intergovernmental Panel on Climate Change (IPCC), reports from 1990, 1995, 2001 and 2007, New York: Cambridge University Press.

Idso, Keith & Craig, 2004, "The Canary in the Coal Mine", *Science*, v. 7: no. 10.

Idso, Sherwood B., 2001, "Carbon Dioxide Induced Global Warming", *AAPG Studies In Geology #47*: p 317-36.

Jacobs, Bonnie, 2007, Assoc. Prof. Dept. of Geol. Sciences, Southern Methodist University: Personal Communication.

Jacobs, Lewis, 2007, President, Institute for Study of Earth and Man, Southern Methodist University: Personal Communication.

Keppler, F. and Rockmann, T., 2007, "Methane, Plants & Climate Change", *Scientific American*: Feb. 2007, p 52-57.

Khilyuk, L. F. and Chilingar, G. V., 2003, "Global Warming: Are We Confusing Cause and Effect?", *Energy*; v. 25: p 357-370.

Khilyuk, L.F. and Chilingar, G. V., 2004, "Global Warming and Long Term Climatic Changes: A Progress Report", *Environmental Geology*, v. 46: p 970-79.

Lamb, H. H. 1995, *Climate History & The Modern World*, 2nd Edition, London, U.K.: Routledge, 443 p.

Lean, J., 2004, "Solar Irradiance", NOAA/NGDC.

Linden, Eugene, 2006, *The Winds of Change*, New York: Simon & Schuster.

Mackenzie, Fred T., 1998, *Our Changing Planet*, 2nd Edition, Upper Saddle River, N.J.: Prentice Hall.

Mann, M.E. et al., 1999, "Northern Hemisphere Temperatures During the Past Millennium", *Geophysical Research Letters*, v. 26: p. 759-762.

Mayewski, P. A. and White, F., 2002, *The Ice Chronicles*, Hanover, N.H.: University Press of New England.

McIntyre, Stephen and McKitrick, Ross, 2003, "Correction to the Mann, et al, Proxy Data Base and Northern Hemisphere Average Temperature Series", *Energy and Environment*, v. 14, No. 6: p 751-771.

Merritts, D., et al., 1998, *Environmental Geology*, New York: W. H. Freeman and Company.

Milloy, Steve, 2006, JunkScience.com, © April 14, 2006, personal communication.

Monnin, et al., 2001, "Atmospheric $CO_2$ Concentrations Over the Last Glacial Termination", *Science*, v. 291, No. 5: p. 112-14.

Montanez, I. P., et al., 2007, "$CO_2$ Forced Climate & Vegetation Instability During the Late Paleozoic Deglaciation", *Science*, v. 315: p. 87-91.

Mudlesee, M. et al., 2001, "The Phase Relations: A More Atmospheric $CO_2$, Quaternary", *Science*, Rev. v. 20: p. 583-89.

NASA, 2006, Science @ NASA, http://science.nasa.gov.headlines/y2006/10may_longrange.html?list3134.

*New York Times, Science Times*, 2006, "Ancient Crash, Epic Wave", Nov. 14, 2006.

Pekarek, Alfred, 2001, "Solar Forcing of Earth's Climate", in Gerhard, Lee C., et al., 2001, "Geological Perspectives of Global Climate Change", *AAPG Studies in Geology #47*: p. 19-34.

Petit, J. R., et al., 1999, "Climate & Atmospheric History of the Past 420,000 Years", *Nature*, v. 39: June 3, 1999, p 429-35.

Rahmstorf, Stefan, 2006, *Climate Change Fact Sheet*, www.pik-potsdam.de/~stefan.

Reiter, Paul, 2000, "From Shakespeare to DeFoe: Malaria in England in the Little Ice Age", *Emerging Infectious Diseases* 6: p 1-11.

Reuters, 2007, "Study: Killer Hurricanes Thrived in Cooler Seas".

Robinson, Arthur B., et al., 1998, "Environmental Effect Increased Atmospheric Carbon Dioxide: Petition Project", La Jolla, CA.

Royer, D. L., et al., 2004, "$CO_2$ As A Primary Driver of Phanerozoic Climate", *GSA Today*, v. 14: p. 4-10.

Royer, D. L., 2006, "$CO_2$ Forced Climate Thresholds during the Phanerozoic", *Geochim. Acta* 70: 5665-5674.

Ruddiman, William F., 2000, *Earth's Climate, Past & Future*, New York: W. H. Freeman & Co.

Scotese, C. R., ©2005, Plate Tectonic and Paleographic maps and animations, PALEOMAP Project (www.scotese.com)

Shaviv, N. J. & Veizer, J., 2003, "Celestial Drivers of Phanerozoic Climate?" *GSA Today*, v. 13: no. 7.

Shaviv, N. J., 2003, "Towards a Solution to the Faint Young Sun Paradox", *Journal of Geophysical Research*, v. 108: p 1437.

Siegenthaler, U., et al., 2005, "Stable Carbon Cycle Climate Relationship", *Science*, v. 310, p. 1313-17.

Singer, Fred S. & Avery, D. T., 2007, *Unstoppable Global Warming*, Lanham, Md.: Rowman, & Littlefield.

Shinn, E. A., 2001, "Coral Reefs & Shoreline Dipsticks", in Gerhard, L. C., et al., *AAPG Studies in Geology #47*, 2001: p. 250-64.

Skinner, Brian J. & Porter, Stephen C., 1995, *The Blue Planet*, New York: John Wiley & Sons, Inc.

Smith, Robert & Siegel, L. J., 2000, *Windows Into The Earth*, New York, Oxford University Press.

Solanki, S. K., et al., 2004, "Unusual Activity of the Sun During Recent Decades Compared to the Previous 11,000 Years", *Nature*, v. 431: p 1084-87.

Soon, W. S. et al., 2001, "Modeling Climatic Effects of Anthropogenic Carbon Dioxide Emissions: Unknowns and Uncertainties", *Climate Research*, v. 18: p 259-275.

Soon, W., et al., 2003, "Reconstructing Climate & Environmental Changes of the Past 1000 Years: a Reappraisal", *Energy & Environment*, v. 14: p 233-96.

Svensmark, Henrik, 2007, "Cosmoclimatology: A New Theory Emerges", *A & G*, v. 48, p.18-24.

Svensmark, H. & Calder, N., 2007, *The Chilling Stars*, Chambridge, U.K., Icon Books, Ltd.

Tabor, Neil, 2007, Assc. Prof. Department of Geological Sciences, Southern Methodist University, personal communication.

Thomasson, M. Ray, 2007, Tomasson & Associates, Denver, Colorado, personal communication.

Usoskin, Schussler, Solanki and Murssaula, 2006, "Solar Activity, Cosmic Rays and Earth's Temperature: a Millennium-Scale Comparison", *J. of Geophysical Res.*, v. 110: A 10102.

Vail, P. R., et al., 1977, "Seismic Stratigraphy and Global Changes of Sea Level", *AAPG Memoir* 26: p 83-87.

Vakulenko, N. V., et al., 2004, "Evidence for the Leading Role of Temperature", *Dolk. Russian Academy of Science, Earth Sciences*, v. 397: p 663-67.

Von Frese, Ralph, 2006, "Big Bang in Antarctica – Killer Crater Found Under Ice", *Ohio State Research News*, Columbus, Ohio.

Ward, Peter Douglas, 2006, *Out of Thin Air*, Washington, D.C., Joseph Henry Press.

Winchester, Simon, 2001, *The Map That Changed the World*, New York: Harper Collins.

Yapp, Crayton, 2004, "$Fe(CO_3)OH$ in Geothite from a Mid-latitude North American Oxisol", *Geochemica & Cosmochemica*, v. 68: no. 5, p 935-47.

Yapp, Creighton, 2005-07, Prof. Dept. of Geological Sciences, Southern Methodist University, Personal Communication.

Zachos, James, et al., 2001, "Trends, Rhythms and Aberrations in Global Climate 65 Ma to Present", *Science*, v. 292: P 686-93.

Zahn, Rainer, 2002, "Milankovitch and Climate: The Orbital Code of Climate Change", *JOIDES Journal*, v 28: p 17-22.

# INDEX

CPSIA information can be obtained
at www.ICGtesting.com
Printed in the USA
LVOW03s1442150916

504774LV00001B/11/P